Hugo F. Franzen

Physical Chemistry of Inorganic Crystalline Solids

With 90 Figures

Springer-Verlag Berlin Heidelberg NewYork
London Paris Tokyo

Professor Hugo Friedrich Franzen

Ames Laboratory-DOE* and Department of Chemistry
Energy & Mineral Resources Research Institute
Iowa State University
Ames, IA 50011/USA

* Operated for the U.S. Department of Energy by Iowa State University under contract No. W-7405-Eng-82. This research was supported by the Office of Basic Energy Sciences, Material Sciences Division.

ISBN 3-540-16580-0 Springer-Verlag Berlin Heidelberg New York
ISBN 0-387-16580-0 Springer-Verlag New York Heidelberg Berlin

Library of Congress Cataloging in Publication Data:
Franzen, H. F. (Hugo Friedrich), 1934
Physical chemistry of inorganic crystalline solids.
Bibliography: p. Includes index.
1. Crystallography. I. Title.
QD905.2.F73 1986 548'.3 86-6437
ISBN 0-387-16580-0 (U.S.)

Printing: H. Heenemann GmbH & Co, Berlin
Bookbinding: Schöneberger Buchbinderei, Berlin
2152/3020-543210

Preface

The field of Physical Chemistry has developed through the application of theories and concepts developed by physicists to properties or processes of interest to chemists. Physicists, being principally concerned with the basic ideas, have generally restricted their attention to the simplest systems to which the concepts applied, and the task of applying the techniques and theories to the myriad substances and processes that comprise chemistry has been that of the physical chemists.

The field of Solid State Chemistry has developed with a major impetus from the synthetic chemists who prepared unusual, novel materials with the principal guiding ideas growing out of an understanding of crystal structure and crystal structure relationships. The novel materials that pour forth from this chemical cornucopia cry out for further characterization and interpretation. The major techniques for the characterization and interpretation of crystalline solids have been developed in the fields of Solid State Physics and Crystallography.

Thus, the need arose for expanding the realm of Physical Chemistry from its traditional concern with molecules and their properties and reactions to include the physics and chemistry of crystalline solids. This book deals with the applications of crystallography, group theory and thermodynamics to problems dealing with non-molecular crystalline solids. It includes an introduction to some of the the important elementary crystalline solid materials, and to some of the important structural phenomena of such solids, e.g., displacive phase transitions, order-disorder phase transition, first- and second-order phase transitions, nonstoichiometry, incommensurate structure, crystallographic shear and population waves. These subjects are approached from the thermodynamic and group-theoretical points of view via Landau theory. The subjects of symmetry in solids and the irreducible representations of space groups are thoroughly developed from the beginning. Some related subjects which are treated for the sake of completeness, although many thorough treatments of these subjects exist, are diffraction theory and elements of the free-electron theory of conducting solids.

This text grew out of a series of lectures that was presented to a graduate class in Solid State Chemistry at Iowa State University, to a group of interested solid-state chemists at the Max-Planck-Institute for Solid-State Research in Stuttgart and to a solid-state chemistry class at Arizona State University, the latter two series of lectures occurring while I was on sabbatical leave from Iowa State University during the 1981–82 academic year. A preliminary publication of some of the chapters of this book appeared as a volume in the Lecture Notes in Chemistry series of Springer Verlag entitled, "Phase Transitions and the Irreducible Representation of Space Groups".

Many colleagues and students have made important corrections to this text. In particular I am pleased to acknowledge the numerous contributions of Dr. J. C. W. Folmer, and the patient and careful work of Ms. Shirley Standley who created a legible transcript from my handwritten notes.

Ames, Iowa/USA, June 1986 Hugo F. Franzen

Table of Contents

Abbreviations

ccp: cubic close packing
hcp: hexagonal close packing
irr. rep.: irreducible representation
sym. op.: symmetry operation
wrt: with respect to

Chapter I

Introduction

I.1 Motivation

There is in progress a promising evolution in the conceptual framework that solid-state chemists utilize in their attempts to understand the interrelationship between crystal structure and physical properties of solids on the one hand and the electronic interactions within the solid on the other. As has been true of chemistry generally, this evolution is closely linked to our ability to meaningfully apply the concepts of basic physics, especially the laws of quantum mechanics, to chemically useful descriptions of real systems, and its ultimate success will hinge upon the essential physical character of the models to which these concepts are applied.

At this point in time a key feature in the unfolding of this conceptual framework is the interplay between experiment and applied theory. In the current development of this application of solid state theory it has been found to be important that calculations be carried out on the important thermodynamic, electrical and structural properties of real solids. As a consequence new uses of theory in the interpretation of solid-state chemistry have been motivated and guided in part by the discovery of new compounds and structures. In turn, many of the developments in the application of theory have been made possible through the development of large scale computational capabilities and, most importantly, through the insights and physical understanding of theoretical physists and chemists. However, the fundamental role of interplay between experiment and theory cannot be exaggerated.

The development under discussion are in the direction of a computational capability to calculate the principal thermodynamic and electrical properties starting from nothing more than a description of the chemical content of the system. One greatly desired by-product of this effort is a set of generalizations in terms of atomic properties (valence, electronic structure) of tendencies for certain structural features to occur. Since the earliest days of solid-state chemistry rationalizations in terms of atomic size and in terms of ionic charge and Madelung constants have been provided for the occurrence of particular structural features. There later appeared models based upon attempts to qualitatively generalize the molecular orbital treatment appropriate to binary molecular compounds and upon the utilization of promotion energies to atomic excited state configurations and their correlation to tendencies to form particular structures. Some simple examples of these approaches will be briefly described below.

Among the developments that have been brought into focus by some recent calculations, the results of which will be discussed in the last chapter, there are a number of criticisms of the traditional models that emerge from a consideration of microscopic results. For example, atomic size arguments have largely foundered on the inability of scientists to agree upon sets of radii for the elements, an inability that ultimately rests upon the fact that it has not been possible to define, at the microscopic level, what is meant by the size of an atom in a solid. In terms of microscopic theory this fact can be seen to rest upon 3 features on the electron density distributions in solids; namely:

1. the radial parts of the electron density distributions of the valence electrons vary from compound to compound, and thus radii must be made compound specific,
2. the valence electron distribution need not be spherically symmetric and thus no single radius defining sphere is appropriate, and
3. even if these relatively fine points are ignored, there appear to be different characteristic distances for the interaction (e.g., in a binary compound) of an atom with a like atom and with an atom of a different element, and it appears that these distances vary with stoichiometry (e.g., interatomic distances are different in Ti_2S [1] with strong Ti—Ti interactions, in TiS [2] with, principally, Ti—S interactions, and weak Ti—Ti interactions, and in TiS_3 [3] with strong S—S interactions (in the direction of S_2^{2-}).

I.2 Brief Overview of Traditional Solid-State Concepts

In spite of these problems consistent sets of tabulated atomic and ionic radii have been very useful in solid-state chemistry, especially in providing guidance in interpreting trends and substitutions in series of structure types. Nonetheless, it is by now clear that consideration of radii cannot provide a basis for a fundamental understanding of crystal structure.

The conceptualization of solids as arrangements of ions packed so as to minimize the energy of the coulombic interactions hat not proved to be generally as useful as was once hoped, in large part because of our inability to meaningfully assign a charge to an ion. It has been thought by numerous scientists that it is possible to consider a solid such as NaCl to be, in fact, Na^+ and Cl^-, charged spheres packed in the well-known rock-salt type structure (Fig. I.1). However, it is now recognized that this ionic model is a point of view, a way of visualizing the solid which yields some understanding of some features of the solid, but that it is conceptually impossible to differentiate this model from that which results from

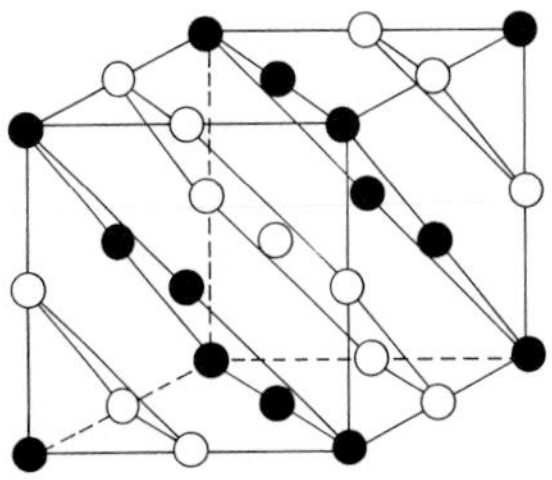

Fig. I.1. NaCl-type structure emphasizing the AbCaBc . . . stacking of alternate hexagonal layers of metals and nonmetals (as described in Chapter IV)

similarly packing Na and Cl atoms [4] allowing for overlap of the incompletely filled Na 2s and Cl 2p atomic orbitals until the resultant molecular (crystal) states are filled. A basic reason for this impossibility is the fact that the sodium to chlorine interatomic spacing can be conceived as a Na^+—Cl^--distance or as a Na—Cl distance simply by taking the radii of defined spheres to be of such magnitudes that either the spheres contain the net charges appropriate to the ions or, alternatively, contain no net charges. In other words the partitioning of electronic charge into spheres surrounding atomic positions is arbitrary, and can lead for the same solid to an ionic or a nonionic interpretation without distinction. The concept of ionic character in a solid is useful to describe electrical conductivity, dissolution in water, optical absorption, etc. However, as has always been known but has sometimes been forgotten, electronic distinctions between what are phenomenologically ionic and covalent solids are not easy to discern. In any case, there is no categorical distinction, nor are there clear-cut microscopic concepts for what is meant by a given fractional ionic character.

It should not be overlooked that the ionic model has been remarkably successful in the calculation of cohesive energies of a number of solids, in particular the alkali halides and the alkaline earth chalcides, however, this success is seldom at a chemically significant level of accuracy (for example about ± 10 kJ per gram atom or better at 1000 K), and since reasons for its relative success or failure (usually ascribed to microscopically rather poorly defined repulsive and polarization terms) are not well-known, it is not possible to predict a priori which calculations will, and which will not, yield useful thermodynamic quantities.

With regard to the chemically more intriguing molecular orbital (Fig. I.2) and valence bond models the problems are more subtle. Molecular orbital modeling to provide a qualitative understanding of the properties and structures of crystalline solids once they are known has been highly successful. Many solids such as the oxides and sulfides of the transition metals, for example, in well known structure types (rock-salt, NiAs, perovskite, $CdCl_2$, etc.) have been well described at the

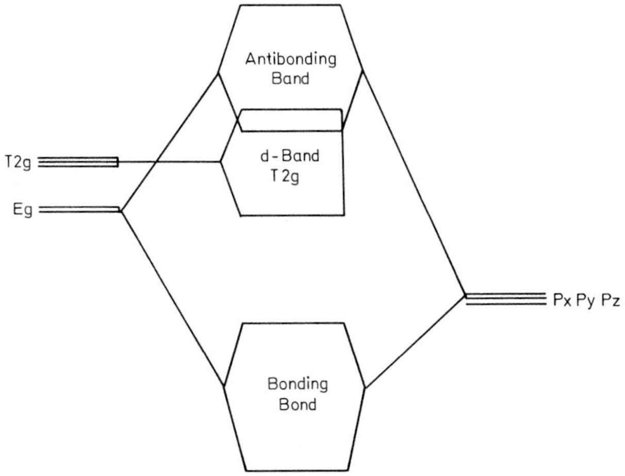

Fig. I.2. A schematic M.O. diagram for transition metal binary compounds

electronic level by molecular orbital models [5]. In fact it appears that this success has in part stimulated the computational theories, and these quantitative theories have in turn encouraged refinement and modification of the qualitative molecular orbital approaches. The principal weakness of the descriptive molecular orbital models lies in their qualitative character and this weakness is currently yielding to the quantitative theories which have the virtue of producing results (electronic configurations) which can be altered only by a new computational effort with a, hopefully, improved approach to the calculation, thus providing a mechanism for advancement. One important residual value of the early qualitative molecular-orbital approaches is that they have provided language (overlap; sigma, pi; bonding, antibonding; etc.) for the expression of the chemical meaning of the results of a calculation.

Problems of appropriately defining the generalizations of molecular orbital based concepts appropriate to 3 dimensional solids do remain, however. For example, the concept of the chemical bond, which has obviously served molecular chemistry extremely well, is not applicable in infinite nonmolecular solids in a straightforward and obvious fashion. For a simple example, consider solid sodium in the body-centered cubic (bcc) or hexagonal close-packed (hcp) structures (both are known for Na(s)). The valence electrons in Na(s) are dispersed over states in a range of energies (a band) with wave functions which are delocalized over a whole crystal domain. It should be remarked that this delocalization is a necessary but not a sufficient condition for metallic conductivity, which arises when the electrons are in states which are in what is essentially a continuum of delocalized states with some neighboring states being unoccupied. This is the case for Na(s) for those electrons in the energy interval within $\sim$kT of the top of the highest filled states, and this amounts at 300 K to only on the order of 0.5–1 % of the total number of the electrons which are delocalized in the quantum mechanical sence [6].

From the description of the delocalized wave functions it is difficult to extract a qualitative picture of a chemical bond. By calculating the electron density distribution ($\Sigma\psi_i\psi_i^*$ as a function of position) it is possible, by comparison with atomic Na(g), to obtain a measure of the change in the electron distribution accompanying the formation of the solid. However, it should be noted that, in the cubic case, for example, the principal contribution to the wave function will be (considering the wave function to be made up of spherical harmonics, which has been found to be useful) s-type, and the next most important will be p-type. Since (as will be discussed) in cubic symmetry P_x, P_y and P_z are equivalent by symmetry, it follows that the only departure from spherical symmetry in the electron density distribution will result from the very small d-type contributions where, because of the familiar $e_g - t_{2g}$ splitting, not all d-types states are equally occupied. As a consequence, to a high degree of precision, the electron-density distribution in bcc Na(s) is a collection of overlapping spheres, i.e., "nondirectional". It is clear that there is a "bonding" effect, namely the solid is energetically more stable than the vapor by the amount of the heat of vaporization, however it is also clear that is in a number of senses inappropriate to call the interactions Na—Na bonds (compare with Na_2(g)) and entirely reasonable to call them Na-crystal effects, i.e., the nature of "bonds" between nearest or next-nearest neighbours (and etc.) is difficult to conceptualize.

Thus the idea of a bond as a localized pariwise effect between atoms in close proximity has little justification for those electron distributions arising from states best described theoretically in the band picture, for which the single electron wave functions are delocalized over the whole solid. It is possible to maintain two quantitative aspects of chemical bonding in the consideration of solids, namely:

1. the total bond energy (i.e., the enthalpy of atomization at 0 K, and
2. the electron density distribution, which can be subdivided into a radial part and an angular part, i.e., the deviation from spherical symmetry of the electron density distribution within a sphere surrounding an atomic position (however, once again, the radius of the sphere is arbitrary).

Thus two characteristics of chemical bonds, the bond energy and the directional nature of the bond, can be discussed in solids without implying that the interactions are pairwise between near neighbors.

I.3 Models of Interactions in Solids

In the case of molecular chemistry the model of hybrid orbitals is frequently used to discuss the directional nature of chemical bonding. A simple extension to solids that is frequently cited is sp^3 hybridization where promotion to the sp^3 electronic configuration is invoked to explain the tetrahedral structure of diamond or silicon. This line of thinking has been extended, for example, by Altmann, Coulson and Hume-Rothery [7] to cases involving more than one ligand per orbital, e.g., the association of sd^3 hydrids (two ligands per orbital) with the bcc structure, and extensively by Engel [8] and Brewer [9] to the general correlation of metal and alloy structures with electronic configurations of low-lying excited states of the gaseous atoms.

However, a difficulty with hybridization schemes, in fact one which makes the correlation a correlation as opposed to a theory, is that it leaves the exact connection between the configurations of low-lying excited states of gaseous atoms and the geometry in the crystalline solid unspecified, i.e., we lack the ability to state exactly what the meaning of a hybrid configuration is in a real solid or molecule. For example, while it is recognized that low-lying excited states with the sp^3 configuration are correlated with tetrahedral structures, it is also true that the photoelectron spectra of tetrahedral molecules and solids do not show the carbon or silicon valence level to be composed of a single state (molecules) or band (solids). For example, for Si(s) there appear bands of states which have been identified by comparison with theory but which are not at all recognizable as (Fig. I.3) a single sp^3 hybrid valence band. Any attempt to bring this fact and the sp^3 hybridization model into consistency results in a molecular-orbital type of interpretation, and not in an explanation which adheres rigorously to the original hybridization concept.

The success of the valence bond concepts in rationalizing and even predicting structures is therefore seen to rest, rather than upon a fundamental correctness of the underlying physical picture, instead upon an ability of the hybridization schemes

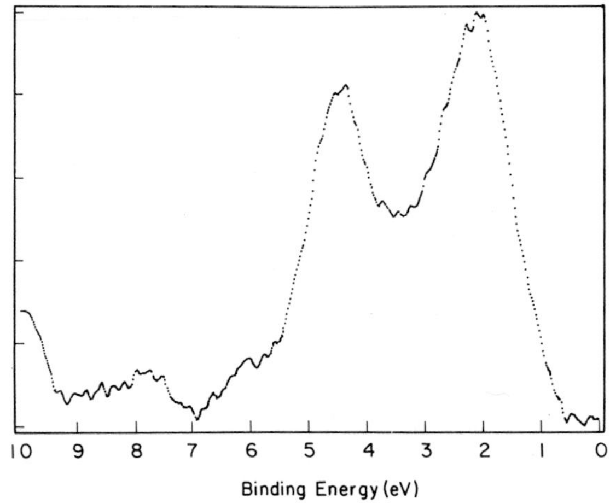

Fig. I.3. The X-ray photoelectron spectrum of Si(s) showing the s-p separation in the valence region

to catch the essence of the electronic factors influencing crystal structure, in spite of a wide disparity between the energetics of the hybrid orbital scheme and those of the real solid.

Were the scope of chemist's interests encompassed by the solids Al, Si, NaCl and TiO, as examples of prototypical metallic, covalent, ionic and mixed compounds, we could, by picking and choosing among the various models discussed above, produce some rather satisfying interrelationships between crystal structure, thermodynamic stability, physical properties and electronic structure.

For Al (s) the Brewer-Engel correlation provides an excellent basis for the discussion of structure and stability. It was first noted by Engel [8] and later powerfully, with numerous applications to alloys and intermetallic compounds, reasserted by Brewer [9] that there is a correlation of structure type (particularly bcc, ccp and hcp and related structures) with the number of s- and p-type electrons in the low-lying configurations of the gaseous atoms. The idea, which is made quantitative in Brewer's papers [9], is that the bond energy (different for different types of electrons) gained from the stable configurations more than compensates for the energy required to promote the atom to an excited configuration. What are then "correlated" are structural features and the number of s and p electrons in the electronic configuration that provides the most stable bonding arrangement. Roughly speaking, sd^{n-1} (n = the number of valence electrons) correlates with bcc, $sd^{n-2}p$ with hcp and $sd^{n-3}p^2$ with ccp, and thus for Al (sp^2 configuration in the "valence excited state") the expected structure is that observed, namely ccp.

The case of Si(s) has been dealt with above. The sp^3 hybrids (which again come into play because the bond-energy gain exceeds the promotion energy loss) yield tetrahedral symmetry. The resulting four bonds directed to four nearest neighbors are just filled by the four valence electrons per Si. Hence the crystal structure and semiconducting behavior are clearly brought into line with the electronic structure.

For the case of NaCl(s), a typical "ionic" compound, the Born-Mayer equation* yields a lattice energy of 749.8 kJ mol^{-1} whereas the value obtained from a Born cycle is 769.0 kJ mol^{-1}, and thus the ionic model provides a good estimate (just within ± 10 kJ per gram atom) for the lattice energy. Furthermore, the sum of the ionic radii ($r_{Na^+} + r_{Cl^-} = 276$ pm [10]) is in very good agreement with the experimental value (281 pm). And, finally, ionic calculations for hypothetical NaCl(s) with other structure types (NiAs, ZnS, etc., but not CsCl) yield higher energies and thus nearly correctly predict the structure type for this solid. Thus the ionic model is quite successful in providing an understanding of NaCl.

In the case of TiO(s), which also crystallizes in the rock-salt type structure, a typical qualitative molecular orbital scheme appears in Fig. I.2. According to this cheme the $e_g(d_{x^2-y^2}, d_{z^2})$ and $t_{2g}(d_{xy}, d_{yz}, d_{xz})$ type atomic orbitals are split by the octahedral field in which Ti is located. The e_g orbitals in part form bonding molecular orbitals with the oxygen p orbitals and are lowered in energy, and in part are increased in energy as a result of the antibonding interaction. The result is a set of valence states which, because of the symmetry properties of solids form a quasicontinuum of states (to be discussed in Chap. X) called the valence band, and a set of higher lying states called the conduction band. The remaining Ti d-type states form a d-band consisting principally of overlapping Ti t_{2g} orbitals. The electrons are distributed among the states such that the valence band is filled, and the d-band is partially filled, giving rise to the observed metallic conduction of TiO. The principal differences between NaCl and TiO (neglecting vacancies), both with the rock-salt type structure, arise, in this view, because of the d-orbitals of Ti which mix to some extent with the oxygen p-type orbitals to give rise to a partial covalent character, and which interact (t_{2g}-type) to form a d-band which gives rise to a metallic character and metallic conductivity.

Thus we have seen from the examples presented above that solid-state chemists are faced with the following situation: although for each of the models there are involved concepts that are ambiguous at the fundamental microscopic level (hybridization, ionic charge, ionic size, molecular interaction) nonetheless by not insisting upon conceptual precision we gain insight that is useful in dealing with typical categories of materials. A critical point is that not all materials fall so neatly into the categories discussed above. It is now known because of the preparation of a host of interesting new materials the stoichiometries, stabilities, structures and properties of which fall almost totally outside of the realm of the models discussed above, that alternative approaches to the consideration of solid compounds must be developed.

A major early development in the application of chemical bonding concepts to solid compounds with metallic character occurred in 1948 when Rundle published an article entitled, "A New Interpretation of Interstitial Compounds.

* $U = \frac{-AZ_iZ_je^2N}{4\pi E_0 r}(1 - \varrho/r)$ where A = 1.74756 = the Madelung constant for the rock-salt type structure, $e = 1.602 \times 10^{-19}$ C, the electronic charge, $E_0 = 8.854 \times 10^{-12}\ J^{-1}\ C^2\ m^{-1}$, r = the anion-cation internuclear distance = 282 pm for NaCl and ϱ = a distance characteristic of anion-cation repulsion = 32.1 pm for NaCl.

Metallic Carbides, Nitrides and Oxides of Composition MX" in the first volume of Acta Cryst. [11]. In this article Rundle questioned the conceptualization of transition-metal carbides, nitrides and oxides as interstitial compounds and pointed out that the 3-center bonding concept could be generalized to include the —M—X—M—X—M—X—. . . strings in the rock-salt type solids. The idea, which was later also used by Slater [4] to advance the view that on the microscopic level the distinction between ionic and covalent bonding in solids is arbitrary, is that the metal bonding orbitals equally overlap both lobes of each of the 3 nonmetal p-orbitals leading naturally to the observed octahedral coordination. Two important points made in Rundle's paper are,

1. it is physically unrealistic and conceptually unproductive to consider solid compounds (e.g., TiO) to occur as the result of hard-sphere oxygen atoms residing in the interstices of hard-sphere, close-packed metal atoms, and
2. it is useful to consider the metal-nonmetal delocalized interactions involving bonds of order less than one (i.e., the appropriate consideration of such interactions could lead to a fundamental understanding of the symmetry, electrical properties and thermodynamic stability of such solids).

Since the time of the appearance of that article there has been a succession of papers on the band theory of TiO (and related compounds) starting with a 1958 paper by Bilz [12] and terminating with a modern, self-consistent A.P.W. calculation by Neckel, Schwarz, Eibler, Weinberger and Rastl in 1975 [13].

A major conclusion from the results of the band theory calculations is that the best picture is one which includes consideration of both the metal-nonmetal interactions and the metal-metal interaction much as they are shown in the M.O. picture of Fig. 1.2. The principal advantages of the band theory results over the other models previously discussed are,

1. the descriptive results of band theory include as special cases generalization of the other models (i.e., the orbitals can be described as combinations of spherical harmonics (hybrids), the electron transfer from metal to nonmetal (ionicity) can be specified (with respect to some arbitrary reference), the metal-metal interactions can be understood (e.g., in terms of states near the Fermi energy, etc.),
2. the theory is adapted to the symmetry of the solid (including translational symmetry, as discussed later) and therefore inherently includes electron delocalization, and
3. the theory is quantitative and therefore the results of a given effort of interpretation are fixed and amenable to improvement, but only through improvement in the initial assumptions.

All of the above are important advantages, and the first two are particularly important for the consideration of solids with a mixture of important metal-metal and metal-nonmetal bonding. While this mixture is now rather well understood in compounds which are related to TiO, this was not always so, e.g., the interstitial model almost completely ignored the M—X interactions and the delocalized multi-centered bonding model of Rundle substantially ignored the M—M interactions. During the time of the development of the more inclusive, and hence more realistic, pictures inherent to the M.O. and band-theories an independent development on the experimental front was fueling the need for the development of such theories.

This experimental development is the unfolding of the richness and variety of solid compounds in which metal-metal interactions are unquestionably important as demonstrated by the structures of the compounds. A whole class of compounds, including halides, chalcides, pnictides and carbides, in which there are portions of the structure in which the metal-metal interactions are comparable to those in the elemental metal have been revealed through the efforts of a number of solid-state chemists, most notably Nowotny [14], Chevrel [15], Corbett [16], McCarley [17] and Simon [18]. The class of compounds also includes the metal-rich sulfides and selenides that have resulted from high-temperature chemical research which have been recently reviewed [19]. From among the compounds in this class a few structures have been chosen as typical yet simple examples for purposes of a brief introduction in Sect. IV.13. It was immediately clear, once the structures of solids such as ScCl [20] and Hf_2S [21] were known, that there was no possibility of understanding the interactions in the solids without taking into account the obviously important metal-metal bonds. Attempts at an ionic interpretation lead immediately to serious problems, namely, for example, the implausibility of +1 cations for the group IV transition metals (given the rest of their known chemistry) and the clear Madelung instability that would result from adjacent metal (cation) layers, not to mention the difficulty of rationalizing the metal-metal distances in terms of some unknown radius characteristic of M^{+1}. Thus it is obvious at the outset, forced by the structures of the solids alone, that a theoretical interpretation of these solids must be in terms of a theory which places a priori the metal-nonmetal und metal-metal interactions on an equal footing, i.e., making no initial assumptions about the relative importance of the interactions.

In this important way the metal-rich solids have contributed to the advance in our understanding of chemistry. Without the emergence of the metal-rich compounds the need for the development of techniques for the application of solid-state theory to intrinsically chemical questions (structure, stability and electronic structure and their interrelationship) could be more easily avoided or postponed. However Hf_2S and ZrCl, and the rich variety of metal-rich solids related to them, require for an understanding of their chemistry the application of theoretical techniques which until quite recently have only been applied to problems in physics and metallurgy.

In 1980 Marchiando, Harmon and Liu [22] reported the elctronic structure of ZrCl as calculated by band theory. They found, in agreement with the observed properties of anisotropic (planar) metallic conductivity and graphitic cleavage, an interaction of Zr-d with Cl-p type states within the Cl—Zr—Zr—Cl layer, and Zr-d interactions within the sheets leading to a 3 eV range of crystal states of mostly Cl-p character and some Zr-d character at a lower energy and, starting about 1 eV higher, a range which is mostly Zr-d in character. This result can be compared with the UPS spectrum of Corbett and Marek [23] shown in Fig. I.4. The charge transfers calculated from the self-consistent band theory results are small (—0.1 el per Zr and +0.2 el per Cl, with the difference being made up by the intersitial regions). Thus band-theory was shown to smoothly provide the overall chemical

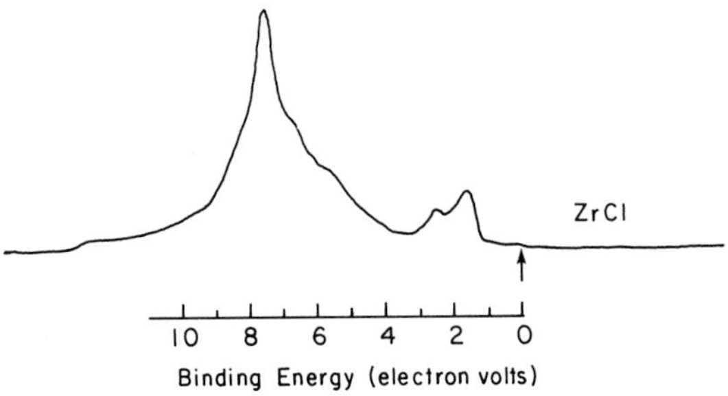

Fig. I.4. The photoelectron spectrum of ZrCl. Zr d-band states are centered at 2 eV, the Cl p-band states at 7.5 eV

bonding picture appropriate to this strongly metal-metal bonded solid for which the structural and bonding models mentioned previously yielded no clear-cut interpretation.

The example of Hf_2S is useful in the discussion of chemistry because it is structurally and chemically very similar to HfS (which, in turn, is "isoelectronic" with TiO). The nearest neighbor environments of sulfur in Hf_2S and HfS are almost identical (both trigonal prisms with very nearly the same S—Hf distance) suggesting that the nature of the S—Hf interactions is not significantly different in the two compounds. In the case of Hf_2S a band-theory approach to the consideration of the interactions is essentially forced by the lack of a physically realistic alternative, in the case of HfS a similar approach would be at least strongly indicated by the similarities between Hf_2S and HfS. Hence the conclusion that band-theory is appropriate to HfS and related MX compounds (NiAs-type, NaCl-type and WC-type), is seen to be a natural extension of the conclusion reached for Hf_2S on the basis of elementary chemical considerations.

It should be noted that the Brewer-Engel correlation has been used to rationalize the difference in structure type between Hf_2S and Hf_2Se on the one hand, and Ti_2S, Ti_2Se, Zr_2S and Zr_2Se on the other [24]. Also it is clear that a molecular orbital model which rationalizes the properties of Hf_2S can be suggested. However neither the valence-bond nor the molecular-orbital model, and certainly not the ionic model, could provide a scheme capable of predicting the existence of a class of stable compounds with metal-metal bonding similar to Hf_2S and ZrCl. The band-theory approach has not yielded such predictions yet either, however it is not unreasonable to hope that total energy calculations will in the future provide a basis for the theoretical consideration of the existence of metal-rich compounds. At any rate, the inability of the older models to deal effectively with the existence and properties of the metal-rich solids has demonstrated a need in the development of the application of theory to solid state chemistry, and it is suggested that the chemist's best hope of meeting this need resides in a return to fundamental theory. Among the advantages of such a return lies the possibility of correctly unraveling the relative importance of the metal-metal interactions and their relationship to the metal-nonmetal interactions via a theory unfettered by initial assumptions which fundamentally encumber the consideration with preconceptions about the electronic state or size of the atoms in a compound.

Band theory of solid-state physics meets all of the above stated requirements, and this theory has in recent years been increasingly applied to materials of chemical interest and, of equal importance to chemists, with an effort made to present chemically useful results. The band-theroy approach derives much of its strength from the fact that the implicit and explicit assumptions upon which it is based are kept to a minimum. The calculations are based upon the fundamental law determining electronic behavior, the Schrödinger equation, subject to the condition (discussed in Chap. X) that the electronic wave functions have the translational and rotational symmetry of the crystals. The most serious assumption is made in constructing a potential in which, it is supposed, individual electrons move. The effects of electron-electron correlation are approximately accounted for by different (e.g., Slater or Hedin-Lundquist exchange) schemes, and it is not possible to determine exactly the effect of this single electron approximation upon the results. However, approximate correction terms are known explicitly and can thus be modified and improved upon in principle.

At any rate a fundamental appreciation of the band-theory approach requires an understanding of solid-state symmetry (translational periodicity and space groups) and it is with the purpose of developing such an understanding that the next chapter initiates a systematic development of the concepts of symmetry in solids. This development proceeds in particular in Chap. III and V with applications following in later chapters.

I.4 Problems

Only brief overviews of a variety of models and correlations were presented in the chapter, therefore additional reading is probably necessary in order to do the following problems. Data from the literature are also required.

1. Discuss the polymorphism of Ti, Zr and Hf from the point of view of the Brewer-Engel correlation.
2. Discuss the structure and electronic behavior of the III–V semiconductors from the point of view of valence bond theory.
3. Calculate the enthalpy of formation of NaCl using the ionic model.
4. Provide a qualitative picture of the bands in ScN with the NaCl-type structure using molecular orbital concepts.
5. Describe why the above approaches are not particularly useful in describing the structure and properties of ScCl.

Chapter II

Space Lattice Symmetry

II.1 Introduction

Some properties of crystalline solids, such as the directions and symmetry of diffraction (X-ray, neutron, electron), the anisotropy of transport phenomena (electrical, thermal conduction, matter diffusion), and the anisotropy of thermal expansion and interactions of crystals with optical radiation are directly related to the underlying three-dimensional periodicities that characterize many crystalline materials. It is therefore important to understand such periodicity in order to better understand the phenomenology of the interactions of crystalline solids. It is also necessary to understand three-dimensional periodicity as a basis for development of the abstract theory of symmetry in crystalline solids (the theory of space groups and their representations). The three-dimensional periodicities are conveniently discussed in terms of space lattices, which are three-dimensional spatial arrays of descrete points which correspond to the translational symmetries of solids. The theory of space lattices, developed below, is the theory of the allowed symmetries **of the periodicity** of three-dimensionally crystalline solids.

II.2 Translational Periodicity

Crystalline solids with **three-dimensional periodicity** are characterized by sets (signified by { }) of **translational symmetry operations,** $\{\mathbf{T}_{mnp}\}$ that are linear combinations of noncoplanar basis vectors, **a**, **b**, **c**:

$$\mathbf{T}_{mnp} = m\mathbf{a} + n\mathbf{b} + p\mathbf{c}\ , \qquad \text{(II.1)}$$

where m, n and p are integers. The lengths of and angles between the basis vectors vary from one solid to another and, subject to restraints to be discussed below, also with the thermodynamic state of a given crystalline solid. The vectors, $\mathbf{T}_{mnp}$, are between all points that have symmetrically equivalent environments.

A point in space within a crystal can be defined relative to an arbitrary origin by $\mathbf{r} + \mathbf{T}_{mnp}$ where

$$\mathbf{r} = x\mathbf{a} + y\mathbf{b} + z\mathbf{c} \qquad \text{(II.2)}$$

with x, y and z between zero and one. All such points with different integral values of m, n and p and fixed x, y and z are points which are indistinguishable by virtue of their environments. The point defined by **r** is frequently designated by x, y, z and the vector $\mathbf{T}_{mnp}$ by m, n, p and the above can be restated, "the points x, y, z and x + m, y + n, z + p are equivalent by symmetry in a crystalline solid exhibiting three-dimensional periodicity".

The set of all points generated by the termination of the vectors $\mathbf{r} + \mathbf{T}_{mnp}$ when m, n and p take on all integral values for a given x, y, z is called a **space lattice.** The space lattice has the property that each lattice point is in a symmetrically equivalent environment. The origin of the space lattice is the point x, y, z, and this point can be chosen with any triple of values between zero and one, i.e., at any point within a **unit cell.**

The three vectors **a**, **b** and **c** thus define a unit cell of the lattice; the volume of this cell is given by

$$V_{cell} = \mathbf{a} \cdot \mathbf{b} \times \mathbf{c} . \tag{II.3}$$

This is the volume per lattice point of the lattice or is the **primitive** cell volume. It is sometimes advantageous to designate cells which describe the symmetry of a lattice (to be discussed below), but for which there exist translational symmetry operations that are not included in the set {Tmnp}. Such cells have more than one lattice point per cell and it is understood that other translational symmetry operations are included in the complete set. For example if it is said that a lattice is **body-centered** then it is understood that the tranlational symmetry operations

$$\left\{\mathbf{T}_{mnp} + \frac{\mathbf{a} + \mathbf{b} + \mathbf{c}}{2}\right\} \tag{II.4}$$

must be added to {Tmnp} to obtain the complete set, i.e., each unit cell of the lattice has, as well as a $\frac{1}{8}$ share of each lattice point at each of 8 corners, a lattice point in the center of the body of the cell. Such body-centered cells contain two lattice points per cell volume. It is then also possible to define a primitive cell, for example by letting

$$\mathbf{a}_p = \frac{1}{2}(\mathbf{a} + \mathbf{b} + \mathbf{c}) , \tag{II.5}$$

$$\mathbf{b}_p = \frac{1}{2}(\mathbf{a} + \mathbf{b} - \mathbf{c}) , \tag{II.6}$$

$$\mathbf{c}_p = \frac{1}{2}(\mathbf{a} - \mathbf{b} + \mathbf{c}) , \tag{II.7}$$

and this cell, labeled p, would contain one lattice point per cell. However, the symmetry of the lattice would not generally be fully described by the conventional description of the symmetry of the primitive cell and in this case the centered cell would be chosen. Other centerings, besides bodycentering, are end-centering (e.g. $(\mathbf{a} + \mathbf{b})/2$ for C centering, $((\mathbf{b} + \mathbf{c})/2$ for A centering, etc.) and face centering $((\mathbf{a} + \mathbf{b})/2$ and $(\mathbf{b} + \mathbf{c})/2$ and $(\mathbf{a} + \mathbf{c})/2)$.

One type of symmetry of space lattices is the proper rotational symmetry of the lattice. A lattice has proper rotational symmetry when there exists a line such that rotation by some minimum angle, α, about the line causes the appearance of the lattice to be indistinguishable from what it was prior to the rotation. All rotational symmetry operations about the axis are then rotations by angles $m\alpha$, where m is an integer, and since rotation by 360° about the axis is a symmetry operation it follows that there exists an integer n such that $n\alpha = 360°$. The rotational symmetry **element** (the line about which the lattice is rotated) is then called an n-fold proper axes and the rotational symmetry **operations** corresponding to this axis are $C_n, C_n^2, C_n^3, \ldots, C_n^n = \varepsilon$, where rotation by 360° is symbolized ε because it is equivalent to the identity operation.

For example if $\alpha = 60°$ then the axis is a 6-fold proper axis and the corresponding operations are C_6 (rotation by 60°), $C_6^2 = C_3$ (rotation by 120°), $C_6^3 = C_2$ (rotation by 180°), $C_6^4 = C_3^2$ (rotation by 240°), C_6^5 (rotation by 300°) and $C_6^6 = \varepsilon$. Note that the operations of a 6-fold axis include those of a 3-fold and 2-fold axis, i.e., a 6-fold axis is necessarily also a 3-fold and a 2-fold axis.

In order to discuss the proper rotational symmetries of space lattices it is convenient to intially treat the proper rotational symmetries of plane lattices. Since any two vectors are necessarily coplanar any lattice defined by the vectors $\{\mathbf{T}_{mnp}\}$ can be generated by the integral repetition of the plane lattice defined by $\{\mathbf{T}_{mn}\}$, where $\mathbf{T}_{mn} = m\mathbf{a} + n\mathbf{b}$, according to the stacking vector $\mathbf{c}$. It follows that the proper rotational symmetries allowed for three-dimensional lattices are limited to those allowed for two-dimensional lattices. In order to consider these rotational symmetries it is necessary to make use of the following result: If a plane lattice has an n-fold symmetry axis then it has an n-fold axis through every lattice point. The proof of this statement is illustrated in Fig. II.1. The open circle represents the location of the

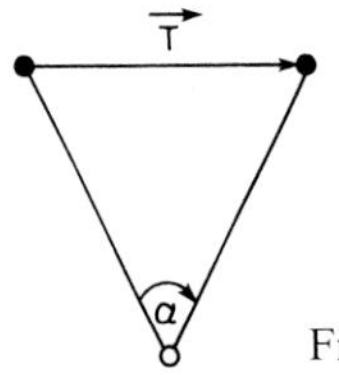

Fig. II.1

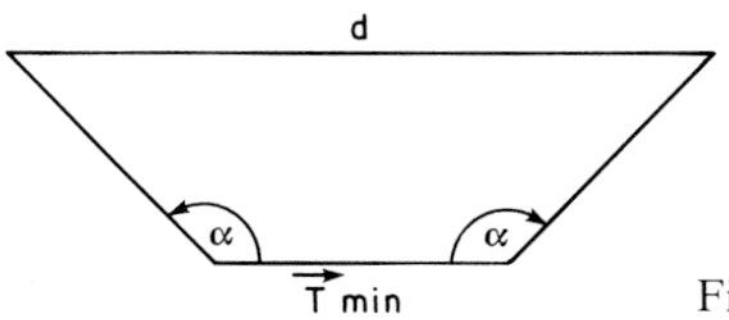

Fig. II.2

Fig. II.1. Open circle: point of intersection of rotational axis (360/α-fold) with plane of figure. Black dots: two lattice points related by rotation by α. **T** is the implied translational symmetry operation

Fig. II.2. Two lattice points related by the shortest vector in the plane ($\mathbf{T}_{min}$) and two additional lattice points (separated by distance d in the direct ion $\mathbf{T}_{min}$) implied by rotation by α and $-\alpha$

symmetry axis-lattice plane intersection. The two black dots represent a lattice point and one related to it by the rotational symmetry operation $C_n \left(n = \frac{360}{\alpha}\right)$.

It follows that **T** is a translational symmetry operation. If, as supposed, C_n is a symmetry operation of the lattice and **T** is therefore also a symmetry operation, then the combined operation $C_n \mid \mathbf{T}$, rotation by α about the open circle followed by translation by **T**, is also a symmetry operation. However $C_n \mid \mathbf{T}$ is equivalent to rotation by α about lattice point, completing the proof.

With this result it can now be shown that translational periodicity significantly restricts the allowed proper rotational symmetries of plane lattices (and therefore of space lattices). Referring to Fig. II.2, we inquire about the allowed values of the length d, the distance between two lattice points which are generated by the proper symmetry operations $\pm C_n$ from two lattice points that are connected by the shortest vector in the plane ($\mathbf{T}_{min}$). The allowed values of d are those which do not imply a translational symmetry operation of length less than $|\mathbf{T}_{min}|$, therefore the allowed values of d are 0, $|\mathbf{T}_{min}|$, 2 $|\mathbf{T}_{min}|$ and 3 $|\mathbf{T}_{min}|$. From these d values the allowed values of α follow, namely 60°, 300°, 90°, 270°, 120°, 240°, 180° and, trivially, 360°. Thus the only allowed axes are 6-fold, 4-fold, 3-fold, 2-fold and, of course, 1-fold. All plane lattices exhibit 2-fold symmetry (both **T** and $-\mathbf{T}$ are translational symmetry operations), it therefore follows that if a plane lattice has 3-fold symmetry it has both 3-fold and 2-fold symmetry through the lattice points, and therefore has 6-fold rotational symmetry. Thus the allowed proper rotational axes through the lattice points of the plane lattices are 2-fold, 4-fold and 6-fold axes.

Another symmetry of plane lattices that can be readily recognized is reflection through a plane perpendicular to the plane lattice. Recognizing the proper rotational

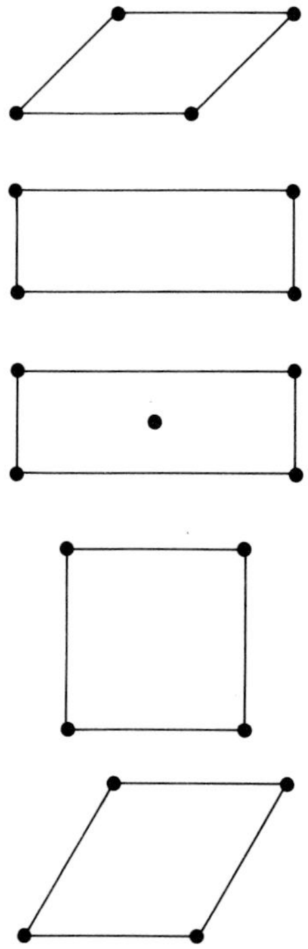

Fig. II.3. The five plane lattice types

symmetries and the reflection symmetries allows the characterization of the allowed plane lattices shown in Fig. II.3. Except for the centered rectangular lattice, the lattices shown follow directly from the proceding discussion. The centered rectangular plane lattice has a nonprimitive cell (two lattice points per cell) which is used for descriptive purposes to call attention to the symmetry that results when the primitive parallelogram cell fulfills the special condition $|\mathbf{a}| = |\mathbf{b}|$, i.e., a parallelogram lattice with $|\mathbf{a}| = |\mathbf{b}|$ has the same symmetry as does the primitive rectangular lattice.

The notation used to describe the lattice symmetries specifies two m's to indicate that there are two different mirror planes which, although they are not equivalent by symmetry (i.e., no symmetry operation carries one into the other) nonetheless are required to occur together by the combined operation of one vertical mirror and the axis. Figure II.4 illustrates the case for a vertical mirror and a 2-fold axis (rectangular lattice symmetry).

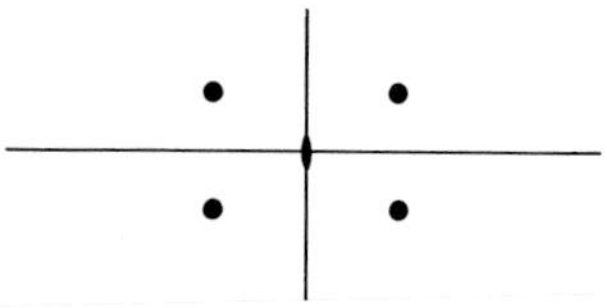

Fig. II.4

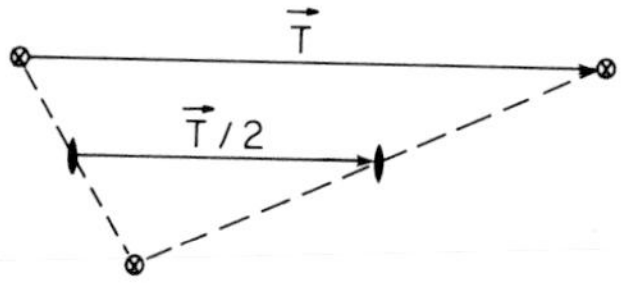

Fig. II.5

Fig. II.4. The four lattice points follow from the two-fold rotation and the reflection operation of one of the mirror planes, the operations of the second mirror plane follow from the combined operations

Fig. II.5. Demonstration that $C_2 \mid \mathbf{T}$ is equivalent to a C_2 about $\mathbf{T}/2$

The lattices shown in Fig. II.3 exhaust the possibilities for plane lattices. This follows from the limitation on the allowed rotational symmetries and from the fact that setting $|\mathbf{a}| = |\mathbf{b}|$ in the rectangular case leads to the square lattice, which is already included. It is natural to inquire why there is no centered square or centered hexagonal lattice. The answers are that a centered square lattice is equivalent to a smaller primitive square lattice, whereas an attempt to construct a centered hexagonal lattice destroys the 6-fold symmetry of the lattice.

II.3 Symmetries of Plane Lattices

As discussed above, all plane lattices have two-fold axes through lattice points and, by their nature, all plane lattices exhibit translational symmetry operations, $\mathbf{T}_{mn}$. The combined operation of a C_2 followed by a translation perpendicular to the axis $(C_2 \mid \mathbf{T}_{mn})$ is illustrated in Fig. II.5. This figure demonstrates that the existence of a 2-fold axis through a lattice point implies the existence of 2-fold axes midway between all pairs of lattice points.

In the case of rotational symmetry other than 2-fold the combination of rotational symmetry operations about a lattice point with translational symmetry operations also results in implied rotational symmetries through points which are not lattice points. However the location and nature of the implied axes must be worked out for each case individually. For the purpose of developing a general scheme for determining the nature and location of implied rotational axes some concepts of operator algebra are useful.

The rotation of a point (x, y) in the a-b plane can be represented by a 2×2 matrix which is symbolized by β:

$$\beta \begin{pmatrix} x \\ y \end{pmatrix} = \begin{pmatrix} x' \\ y' \end{pmatrix}. \tag{II.8}$$

For example, if $\beta = C_{2z}$

$$\begin{pmatrix} \bar{1} & 0 \\ 0 & \bar{1} \end{pmatrix} \begin{pmatrix} x \\ y \end{pmatrix} = \begin{pmatrix} \bar{x} \\ \bar{y} \end{pmatrix}. \tag{II.9}$$

The rotation followed by translation, say by **a**, can be represented by

$$\begin{pmatrix} \bar{1} & 0 \\ 0 & \bar{1} \end{pmatrix} \begin{pmatrix} x \\ y \end{pmatrix} + \begin{pmatrix} 1 \\ 0 \end{pmatrix} = \begin{pmatrix} \bar{x} + 1 \\ \bar{y} \end{pmatrix} \tag{II.10}$$

and, as shown above (Fig. II.5) this operation corresponds to a C_{2z} operation about **a**/2. It is possible to combine the translation and rotation in a single 3×3 matrix:

$$\begin{pmatrix} \bar{1} & 0 & 1 \\ 0 & \bar{1} & 0 \\ 0 & 0 & 1 \end{pmatrix} \begin{pmatrix} x \\ y \\ 1 \end{pmatrix} = \begin{pmatrix} \bar{x} + 1 \\ \bar{y} \\ 1 \end{pmatrix} \tag{II.11}$$

where the last rows of the 3×3 and 3×1 matrices are formally 001 and 1, respectively, so that matrix multiplication yields the correct result. It is also worthwile to note that rotation by C_{2z} about **a**/2 is equivalent to the sequence of operations: translation by −**a**/2, rotation by C_{2z}, translation by **a**/2, i.e., can be represented by

$$\begin{pmatrix} 1 & 0 & \frac{1}{2} \\ 0 & 1 & 0 \\ 0 & 0 & 1 \end{pmatrix} \begin{pmatrix} \bar{1} & 0 & 0 \\ 0 & \bar{1} & 0 \\ 0 & 0 & 1 \end{pmatrix} \begin{pmatrix} 1 & 0 & -\frac{1}{2} \\ 0 & 1 & 0 \\ 0 & 0 & 1 \end{pmatrix}, \tag{II.12}$$

which multiply to yield $\begin{pmatrix} \bar{1} & 0 & 1 \\ 0 & \bar{1} & 0 \\ 0 & 0 & 1 \end{pmatrix}$, as they should.

It was our purpose here to discover a means for determining what any proper rotational symmetry operation through a lattice point of a plane lattice implies regarding other symmetry axes. As a first example we consider the C_4 through the lattice points of the square lattice. Since rotation by 90° about the origin is represented by

$$\begin{pmatrix} 0 & \bar{1} \\ 1 & 0 \end{pmatrix},$$

we can write for a C_4 at $x = t_1$, $y = t_2$,

$$\begin{pmatrix} 1 & 0 & t_1 \\ 0 & 1 & t_2 \\ 0 & 0 & 1 \end{pmatrix} \begin{pmatrix} 0 & \bar{1} & 0 \\ 1 & 0 & 0 \\ 0 & 0 & 1 \end{pmatrix} \begin{pmatrix} 1 & 0 & -t_1 \\ 0 & 1 & -t_2 \\ 0 & 0 & 1 \end{pmatrix} = \begin{pmatrix} 0 & \bar{1} & t_1 + t_2 \\ 1 & 0 & t_2 - t_1 \\ 0 & 0 & 1 \end{pmatrix} \qquad \text{(II.13)}$$

and equate this with the matrix representing a 4-fold rotation operation about the origin plus translation by $m\mathbf{a} + n\mathbf{b}$:

$$\begin{pmatrix} 0 & 1 & t_1 + t_2 \\ 1 & 0 & t_2 - t_1 \\ 0 & 0 & 1 \end{pmatrix} = \begin{pmatrix} 0 & \bar{1} & m \\ 1 & 0 & n \\ 0 & 0 & 1 \end{pmatrix} \qquad \text{(II.14)}$$

i.e., $t_1 + t_2 = m$ and $t_2 - t_1 = n$, relations which fix the location (t_1, t_2) of the axis implied by C_4 through the origin and $\mathbf{T}_{mn}$. For example if translation by $\mathbf{a}$ is chosen then $m = 1$ and $n = 0$ and $t_2 = t_1 = \frac{1}{2}$. It follows that a square lattice necessarily has 4-fold axes through the centers of the unit cells, a fact that is obvious upon inspection of such a lattice.

The same method can be applied to a hexagonal lattice. Rotation by 60° in the hexagonal lattice carries x, y into x-y, x (see Fig. III.1), i.e.,

$$C_6 = \begin{pmatrix} 1 & \bar{1} \\ 1 & 0 \end{pmatrix}$$

and thus $C_3 = C_6^2 = \begin{pmatrix} 1 & \bar{1} \\ 1 & 0 \end{pmatrix} \begin{pmatrix} 1 & \bar{1} \\ 1 & 0 \end{pmatrix} = \begin{pmatrix} 0 & \bar{1} \\ 1 & \bar{1} \end{pmatrix}$

Accordingly we have

$$\begin{pmatrix} 1 & 0 & t_1 \\ 0 & 1 & t_2 \\ 0 & 0 & 1 \end{pmatrix} \begin{pmatrix} 1 & \bar{1} & 0 \\ 1 & 0 & 0 \\ 0 & 0 & 1 \end{pmatrix} \begin{pmatrix} 1 & 0 & \bar{t}_2 \\ 0 & 1 & \bar{t}_1 \\ 0 & 0 & 1 \end{pmatrix} = \begin{pmatrix} 1 & \bar{1} & 1 \\ 1 & 0 & 0 \\ 0 & 0 & 1 \end{pmatrix} \qquad \text{(II.15)}$$

and

$$\begin{pmatrix} 1 & 0 & t'_1 \\ 0 & 1 & t'_2 \\ 0 & 0 & 1 \end{pmatrix} \begin{pmatrix} 1 & \bar{1} & 0 \\ 1 & \bar{1} & 0 \\ 0 & 0 & 1 \end{pmatrix} \begin{pmatrix} 1 & 0 & \bar{t}'_1 \\ 0 & 1 & \bar{t}'_2 \\ 0 & 0 & 1 \end{pmatrix} = \begin{pmatrix} 0 & \bar{1} & 1 \\ 1 & \bar{1} & 0 \\ 0 & 0 & 1 \end{pmatrix} \qquad \text{(II.16)}$$

for rotation, respectively, by 60° and 120° followed by translation by **a**. By matrix multiplication

$$\begin{pmatrix} 1 & \bar{1} & t_2 \\ 1 & 0 & t_2 - t_1 \\ 0 & 0 & 1 \end{pmatrix} = \begin{pmatrix} 1 & \bar{1} & 1 \\ 1 & 0 & 0 \\ 0 & 0 & 1 \end{pmatrix} \tag{II.17}$$

and

$$\begin{pmatrix} 0 & \bar{1} & t_1' + t_2' \\ 1 & \bar{1} & 2t_2' - t_1' \\ 0 & 0 & 1 \end{pmatrix} = \begin{pmatrix} 0 & \bar{1} & 1 \\ 1 & \bar{1} & 0 \\ 0 & 0 & 1 \end{pmatrix} \tag{II.18}$$

from which $t_1 = t_2 = 1$ (the implied 60° rotation is about the lattice point at $x = y = 1$) and $t_1' = \frac{2}{3}$, $t_2' = \frac{1}{3}$ (the implied 120° rotation is about the point $x = \frac{2}{3}$, $y = \frac{1}{3}$). The resulting symmetry axes are diagrammed for all plane lattices in Fig. II.6.

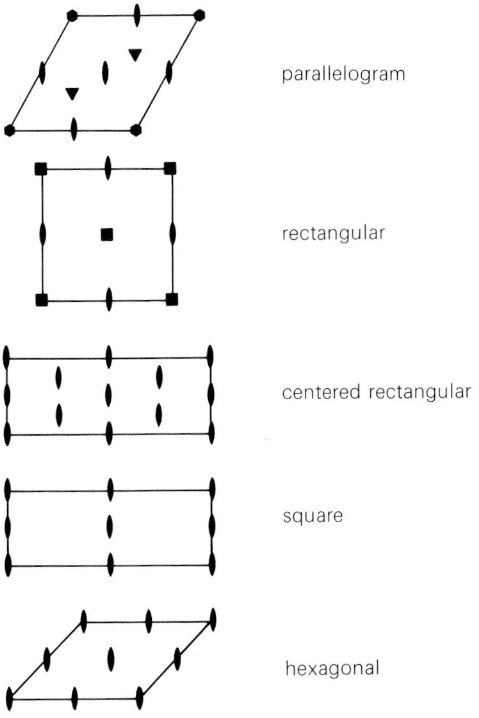

Fig. II.6. The rotational symmetry axes of the plane lattice types

II.4 Space Lattices From Stacking of Plane Lattices

With the knowledge of the plane lattice types and their proper rotational symmetries the possible space lattices can be obtained by considering stacking of the plane lattices with various choices of stacking vector, **c**. As with the plane lattices, the space lattices are characterized according to their symmetries. Therefore the approach to the consideration of allowed space lattice types is to

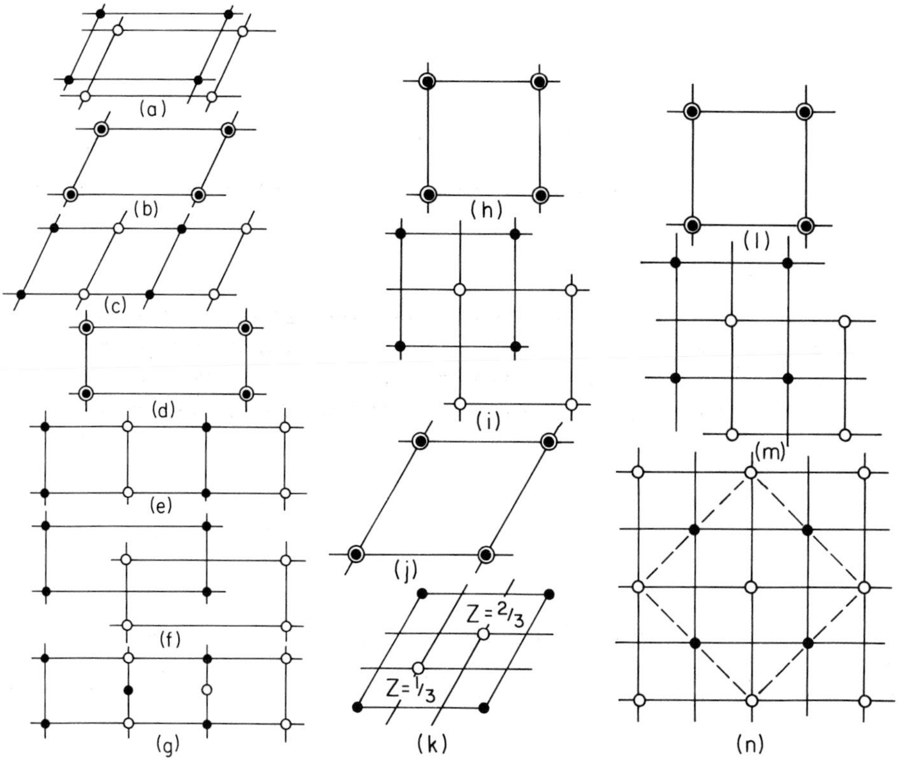

Fig. II.7 a—n. The 14 Bravais Lattice Types.

a triclinic (1) $|\mathbf{a}| \neq |\mathbf{b}| \neq |\mathbf{c}|$, $\alpha \neq \beta \neq \gamma$
b monoclinic (2/m) $|\mathbf{a}| \neq |\mathbf{b}| \neq |\mathbf{c}|$, ($\alpha = \gamma = 90°$, $\beta \neq 90°$, **b** unique)
c end-centered monoclinic (2/m)
d orthorhombic (mmm) $|\vec{a}| \neq |\vec{b}| \neq \vec{c}|$, $\alpha = \beta = \gamma = 90°$
e end-centered orthorhombic (mm)
f body-centered orthorhombic (mmm)
g face-centered orthorhombic (mmm)
h tetragonal (4/mmm) $|\mathbf{a}| = |\mathbf{b}| \neq |\mathbf{c}|$, $\alpha = \beta = \gamma = 90°$
i body-centered tetragonal (4/mmm)
j hexagonal (6/mmm) $|\mathbf{a}| = |\mathbf{b}| \neq |\mathbf{c}|$, $\alpha = \beta = 90°$, $\gamma = 120°$
k rhombohedral (3m) (shown as "doubly centered" hexagonal)
l cubic (m3m) $|\mathbf{a}| = |\mathbf{b}| = |\mathbf{c}|$, $\alpha = \beta = \gamma = 90°$
m body-centered cubic (m3m)
n face-centered cubic (m3m)

examine the ways in which the plane lattices can be stacked so as to preserve axial symmetries. For example, if plane lattices with hexagonal symmetry are stacked such that their 6-fold axes coincide (**c** perpendicular to the a-b plane) then the space lattice has 6-fold symmetry. If the 6-fold axes do not coincide then the 6-fold symmetry is lost, however 3-fold symmetry can be maintained if the 6-fold and 3-fold axes coincide, i.e., if: ($\mathbf{c} = \frac{2}{3}\mathbf{a} + \frac{1}{3}\mathbf{b} + \tau\,(\mathbf{a}\times\mathbf{b})$ or $\mathbf{c} = \frac{1}{3}\mathbf{a} + \frac{2}{3}\mathbf{b} + \tau\,(\mathbf{a}\times\mathbf{b})$ where τ is some scalar fixing the height of the stacking layer). The systematic application of this approach leads to the 14 Bravais lattice types, as follows.

A lattice which exhibits no rotational symmetry is called **triclinic,** and this lattice type can be produced by stacking parallelogram plane lattices with a general **c** (no two-folds coincide). This lattice (Fig. II.7a) has only inversion symmetry (a symmetry of all lattices since if **T** is a translational symmetry operation so too is $-\mathbf{T}$) and in general $|\mathbf{a}| \neq |\mathbf{b}| \neq |\mathbf{c}|$ and $\alpha \neq \beta \neq \gamma$ and all angles differ from 90°. Recalling that we are discussing the symmetry of the periodicity of a crystalline solid, it should be noted that the significance of the inequalities is not that they hold at all thermodynamic states of the solid, but rather that while an equality may occur at some isolated thermodynamic state. When such an **isolated** equality occurs it is a "chance" occurrence at that state and is not generally observed as a result of the symmetry of the system. This remark is generally valid for all inequalities given for the various space lattice types.

There are three stackings which maintain the 2-fold axes of parallelogram plane lattices, one which places the lattice points of the stacked parallelogram lattice directly above those of the basal parallelogram lattice (Fig. II.7b), one which places the lattice points of the stacked lattice above the midpoints of the edges (Fig. II.7c) and one which places the lattice points of the stacked layer above the center of the cell (Fig. II.8). All three of the resulting lattices exhibit 2/m (a two-fold axis perpendicular to a mirror plane) symmetry, a lattice symmetry called **monoclinic.** Figure II.8 shows the equality of the lattice types from

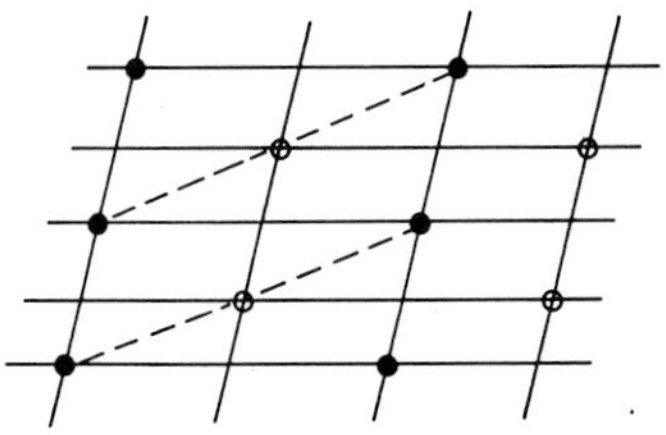

Fig. II.8

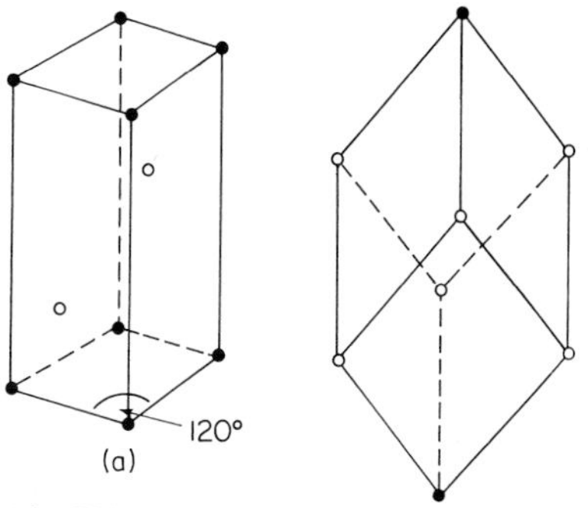

Fig. II.9

Fig. II.8. Stacking of parallelogram plane lattice above base center yields body-centered monoclinic which is equivalent to end-centered monoclinic (dashed lines)

Fig. II.9. (a) Triply primitive hexagonal cell.
(b) Primitive rhombohedral cell

the latter two stackings, i.e., the equivalence of the end-centered and body-centered monoclinic lattice types. The usual convention is to describe this lattice type as end-centered monoclinic. In the monoclinic cases the crystallographic axis along the direction of the 2-fold axes is called the unique axis. It is customary to designate the unique axis to be **b** or **c** with the former choice generally preferred.

With the **b** axis chosen as the unique axis the β angle (the angle between **a** and **b**) is the interaxial angle that differs from 90°. The centered face, when one occurs, is named acording to the crystallographic axis not included in the face, and the usual monoclinic convention is to take this as the C face, i.e., the face formed by **a** and **b**. Thus the rectangular face of the end-centered monoclinic cell is conventionally the a-b face and the translations implied by C-centering are given by $\frac{\mathbf{a} + \mathbf{b}}{2}$ added to all translational symmetry operations implied by **a**, **b**, and **c**.

It is also possible to obtain monoclinic space lattices by stacking rectangular plane lattices, primitive rectangular stacked with **c** projecting onto one edge of the basal unit cell yields primitive monoclinic, and centered rectangular similarly stacked yields an end-centered monoclinic space lattice.

The rectangular lattices can also be stacked to yield space lattices with 2/m 2/m 2/m (three mutually orthogonal mirror planes with three mutually orthogonal two-fold axes along their intersection lines) symmetry in four different ways. The lattice of Fig. II.7d is produced by stacking primitive rectangular lattices, the lattice of Figs. II.7e and II.7f are produced by stacking either primitive or centered rectangular lattices and the lattice of Fig. II.7g is produced by the stacking of centered rectangular lattices. In each case the stacking is done so as to maintain the 2-fold and mirror symmetries of the plane lattices in the resulting space lattices, and a horizontal mirror plane and a horizontal two-fold axis are found in the resulting **orthorhombic** space lattices.

The plane square lattice type can be stacked in five different ways that maintain the 4-fold rotational symmetry in the space lattices. One places the lattice points of the stacked plane directly above those of the basal plane, at an arbitrary height (II.7h) and a second places the lattice points of the stacked plane above the centers of the basal unit cells at an arbitrary height (II.7i). In both of these cases the space lattice has 4/mmm (a four-fold axis with two vertical and one horizontal mirror planes) symmetry and the space lattice types are primitive and body-centered **tetragonal**, respectively. It is left as an exercise to show how a face-centered tetragonal lattice is generated and to demonstrate that the resultant lattice type is equivalent to a body-centered tetragonal space lattice. The tetragonal space lattices are also generated from rectangular plane lattices if a vertical stacking vector is the same length as either **a** or **b**.

When the stacking heights of the square lattices are not arbitrary it is possible to introduce additional symmetry elements. When the vertical stacking of Fig. II.7(l) occurs with $|\mathbf{a}| = |\mathbf{b}| = |\mathbf{c}|$ a simple cubic lattice results. When the stacking of II.7(m) occurs with the stacked layer at height $|\mathbf{a}|/2$ a body-centered cubic lattice results, and when the stacking height is $|\mathbf{a}|/\sqrt{2}$ a face-centered cubic lattice is obtained (Fig. II.7(n)). Note that the resulting cubic symmetry includes

3-fold axes along the body diagonals, and that it therefore follows that cubic lattices could also be generated by the appropriate stacking of hexagonal plane lattices.

The remaining cases are those that result most directly from stacking of plane lattices with 6-fold rotational symmetry. If **c** is perpendicular to **a** and **b** then primitive hexagonal space lattices result. If, on the other hand, the 6-fold axis of the stacked lattice is coincident with the 3-fold axis of the base lattice then the stacking vector must be repeated twice before a complete unit cell of a space lattice with 3-fold symmetry is obtained. The resulting hexagonal cell is triply primitive, and the primitive cell is rhombohedral, as shown in Fig. II.9.

The 14 Bravais lattices are thus generated by the stacking of plane lattices. The fact that there exist no more than 14 space lattice types is made plausible by the limitations placed upon the rotational symmetries of plane lattices by the lattice periodicity and by the limitations placed on the number of ways that plane lattices can stack by the requirement that proper rotational axes superimpose if symmetry is to be maintained in the space lattices.

II.5 Problems

1. Calculate the volumes of the unit cells for
   a. La, hexagonal, a = 3.770, c = 12.159 Å.
   b. S, monoclinic, a = 10.92, b = 10.98, c = 11.04 Å, $\beta = 83°16'$.
   c. As, rhombohedral, a = 4.1318 Å, $\alpha = 54°8'$.
2. Find the a and c axis lengths for the hexagonal cell of As the rhombohedral cell for which is described in 1c.
3. Show that the following plane lattices should not be added to those listed in Fig. II.3.
   a. centered parallelogram.
   b. edge-centered rectangular.
   c. centered hexagonal.
4. Prove, using the appropriate matrices, that a centered plane rectangular lattice has a two-fold axis at $\frac{1}{4}, \frac{1}{4}$.
5. Find the stacking vector **c** that will produce a cubic space lattice by stacking of plane hexagonal lattices.
6. Calculate the density of monoclinic $PdBi_2$ with a = 12.74, b = 4.25, c = 5.665 Å, $\beta = 102°35'$ and 4 formula units per cell.
7. Sketch a projection of a hexagonal space lattice along **c** and show that hexagonal is a special case of end-centered orthorhombic. What is the relationship of a_{ortho} to b_{ortho} if the cell has hexagonal symmetry.
8. Describe how the following distortions might occur continuously
   a. face-centered cubic to body-centered tetragonal.
   b. cubic to rhombohedral.
   c. hexagonal to orthorhombic.
9. Describe the space lattice for the close packed spheres packed ABCBACBCACBACABACBABCBACBCACBACABACB . . .

Chapter III

Space Group Symmetry

III.1 Introduction

Thus far the treatment of symmetry has been restricted to the proper rotational and reflection symmetries of space lattices. The discussion of the symmetry of crystalline solids does not end with the presentation of the 14 Bravais lattice types because the symmetry of a solid is the symmetry of its three-dimensionally periodic particle density, and there are more symmetries available to such periodic patterns than to the lattices which characterize their translational symmetries. This is because of the existence of symmetry operations appropriate to such patterns but not to their lattices, for example the pattern need not be centrosymmetric while the lattice must be.

III.2 Proper and Improper Rotations

The symmetry operations of crystalline solids can be viewed in terms of their effect upon x, y, z in 3-dimensional space. For example, a two-fold axis through the origin and parallel to the **c** axis takes x, y, z into $\bar{x}$, $\bar{y}$, z and thus can be represented by the matrix

$$\begin{pmatrix} \bar{1} & 0 & 0 \\ 0 & \bar{1} & 0 \\ 0 & 0 & 1 \end{pmatrix} \begin{pmatrix} x \\ y \\ z \end{pmatrix} = \begin{pmatrix} \bar{x} \\ \bar{y} \\ z \end{pmatrix}, \qquad \text{(III.1)}$$

and similarly the 3×3 matrices for all the proper rotations of the groups O_h and D_{6h} can be written down (Tables III.1 and III.2). Matrices for all proper rotations of all space groups are contained in these tables. Figure III.1 clarifies the perhaps not immediately obvious relationship of x, y, z and the point related to it by a C_{6z} operation (x — y, x, z) in a hexagonal lattice.

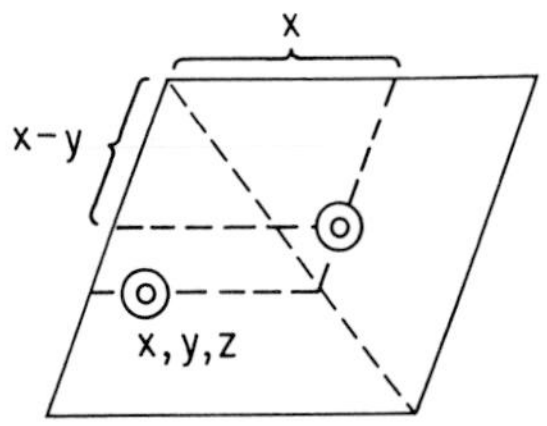

Fig. III.1. Illustration of relationship of positional parameters under C_{6z} in a hexagonal lattice

The proper rotational symmetry operations of a crystalline solid form a closed set (in fact a group). This implies that if two such operations are members of the set then the combined operation is also. For example, in O_h the 90° rotation along z together with the 120° rotation along the body diagonal ($C_{3(x+y+z)}$) combine according to

$$\begin{pmatrix} 0 & \bar{1} & 0 \\ 1 & 0 & 0 \\ 0 & 0 & 1 \end{pmatrix}\begin{pmatrix} 0 & 0 & 1 \\ 1 & 0 & 0 \\ 0 & 1 & 0 \end{pmatrix} = \begin{pmatrix} \bar{1} & 0 & 0 \\ 0 & 0 & 1 \\ 0 & 1 & 0 \end{pmatrix} \tag{III.2}$$

Table III.1. Proper Rotational Symmetry Operations of O_h

$\begin{pmatrix} 1 & 0 & 0 \\ 0 & 1 & 0 \\ 0 & 0 & 1 \end{pmatrix}$	ε	$\begin{pmatrix} 0 & \bar{1} & 0 \\ 0 & 0 & \bar{1} \\ 1 & 0 & 0 \end{pmatrix}$	$C_{3(x-y+z)}$
$\begin{pmatrix} 1 & 0 & 0 \\ 0 & \bar{1} & 0 \\ 0 & 0 & \bar{1} \end{pmatrix}$	C_{2x}	$\begin{pmatrix} 0 & 0 & 1 \\ 1 & 0 & 0 \\ 0 & 1 & 0 \end{pmatrix}$	$C_{3(x+y+z)}$
$\begin{pmatrix} \bar{1} & 0 & 0 \\ 0 & 1 & 0 \\ 0 & 0 & \bar{1} \end{pmatrix}$	C_{2y}	$\begin{pmatrix} 0 & 0 & 1 \\ \bar{1} & 0 & 0 \\ 0 & \bar{1} & 0 \end{pmatrix}$	$C^2_{3(x-y+z)}$
$\begin{pmatrix} \bar{1} & 0 & 0 \\ 0 & \bar{1} & 0 \\ 0 & 0 & 1 \end{pmatrix}$	C_{2z}	$\begin{pmatrix} 0 & 0 & \bar{1} \\ 1 & 0 & 0 \\ 0 & \bar{1} & 0 \end{pmatrix}$	$C^2_{3(x+y-z)}$
$\begin{pmatrix} 0 & 1 & 0 \\ 0 & 0 & 1 \\ 1 & 0 & 0 \end{pmatrix}$	$C^2_{3(x+y+z)}$	$\begin{pmatrix} 0 & 0 & 1 \\ 0 & 1 & 0 \\ \bar{1} & 0 & 0 \end{pmatrix}$	C_{4y}
$\begin{pmatrix} 0 & 1 & 0 \\ 0 & 0 & \bar{1} \\ \bar{1} & 0 & 0 \end{pmatrix}$	$C_{3(x+y-z)}$	$\begin{pmatrix} 0 & 0 & \bar{1} \\ \bar{1} & 0 & 0 \\ 0 & 1 & 0 \end{pmatrix}$	$C^2_{3(-x+y+z)}$
$\begin{pmatrix} 0 & \bar{1} & 0 \\ 0 & 0 & 1 \\ \bar{1} & 0 & 0 \end{pmatrix}$	$C_{3(-x+y+z)}$	$\begin{pmatrix} 0 & \bar{1} & 0 \\ \bar{1} & 0 & 0 \\ 0 & 0 & \bar{1} \end{pmatrix}$	$C_{2(-x+y)}$
$\begin{pmatrix} 0 & \bar{1} & 0 \\ 1 & 0 & 0 \\ 0 & 0 & 1 \end{pmatrix}$	C_{4z}	$\begin{pmatrix} 0 & 0 & \bar{1} \\ 0 & \bar{1} & 0 \\ \bar{1} & 0 & 0 \end{pmatrix}$	$C_{2(-x+z)}$
$\begin{pmatrix} 0 & 1 & 0 \\ \bar{1} & 0 & 0 \\ 0 & 0 & 1 \end{pmatrix}$	C^3_{4z}	$\begin{pmatrix} 0 & 0 & \bar{1} \\ 0 & 1 & 0 \\ 1 & 0 & 0 \end{pmatrix}$	C^3_{4y}
$\begin{pmatrix} 0 & 1 & 0 \\ 1 & 0 & 0 \\ 0 & 0 & \bar{1} \end{pmatrix}$	$C_{2(x+y)}$	$\begin{pmatrix} 0 & 0 & 1 \\ 0 & \bar{1} & 0 \\ 1 & 0 & 0 \end{pmatrix}$	$C_{2(x+z)}$

Table III.1 (continued)

$$\begin{pmatrix} \bar{1} & 0 & 0 \\ 0 & 0 & \bar{1} \\ 0 & \bar{1} & 0 \end{pmatrix} C_{2(-y+z)}$$

$$\begin{pmatrix} \bar{1} & 0 & 0 \\ 0 & 0 & 1 \\ 0 & 1 & 0 \end{pmatrix} C_{2(y+z)}$$

$$\begin{pmatrix} 1 & 0 & 0 \\ 0 & 0 & \bar{1} \\ 0 & 1 & 0 \end{pmatrix} C_{4x}$$

$$\begin{pmatrix} 1 & 0 & 0 \\ 0 & 0 & 1 \\ 0 & \bar{1} & 0 \end{pmatrix} C_{4x}^3$$

and, by inspection of Table III.1, the resultant matrix represents $C_{2(y+z)}$, the 180° rotation about a face diagonal.

In the space groups it is necessary to include operations which are represented by the matrices of Tables III.1 and III.2 with the signs of all nonzero matrix elements changed, i.e., multiplied by the matrix representing inversion. The resultant matrices represent **operations** of rotation followed by inversion (not necessarily the

Table III.2. Proper Symmetry Operations of D_6

$$\begin{pmatrix} 1 & 0 & 0 \\ 0 & 1 & 0 \\ 0 & 0 & 1 \end{pmatrix} \varepsilon \qquad \begin{pmatrix} \bar{1} & 0 & 0 \\ \bar{1} & 1 & 0 \\ 0 & 0 & \bar{1} \end{pmatrix} C_{2y}$$

$$\begin{pmatrix} 1 & \bar{1} & 0 \\ 1 & 0 & 0 \\ 0 & 0 & 1 \end{pmatrix} C_{6z} \qquad \begin{pmatrix} 0 & \bar{1} & 0 \\ \bar{1} & 0 & 0 \\ 0 & 0 & \bar{1} \end{pmatrix} C_{2(x-y)}$$

$$\begin{pmatrix} 0 & \bar{1} & 0 \\ 1 & \bar{1} & 0 \\ 0 & 0 & 1 \end{pmatrix} C_{3z} \qquad \begin{pmatrix} 1 & \bar{1} & 0 \\ 0 & \bar{1} & 0 \\ 0 & 0 & \bar{1} \end{pmatrix} C_{2x}$$

$$\begin{pmatrix} \bar{1} & 0 & 0 \\ 0 & \bar{1} & 0 \\ 0 & 0 & 1 \end{pmatrix} C_{2z} \qquad \begin{pmatrix} 1 & 0 & 0 \\ 1 & \bar{1} & 0 \\ 0 & 0 & \bar{1} \end{pmatrix} C_{2(2x+y)}$$

$$\begin{pmatrix} \bar{1} & 1 & 0 \\ \bar{1} & 0 & 0 \\ 0 & 0 & 1 \end{pmatrix} C_{3z}^2 \qquad \begin{pmatrix} \bar{1} & 1 & 0 \\ 0 & 1 & 0 \\ 0 & 0 & \bar{1} \end{pmatrix} C_{2(x+2y)}$$

$$\begin{pmatrix} 0 & 1 & 0 \\ \bar{1} & 1 & 0 \\ 0 & 0 & 1 \end{pmatrix} C_{6z}^5 \qquad \begin{pmatrix} 0 & 1 & 0 \\ 1 & 0 & 0 \\ 0 & 0 & \bar{1} \end{pmatrix} C_{2(x+y)}$$

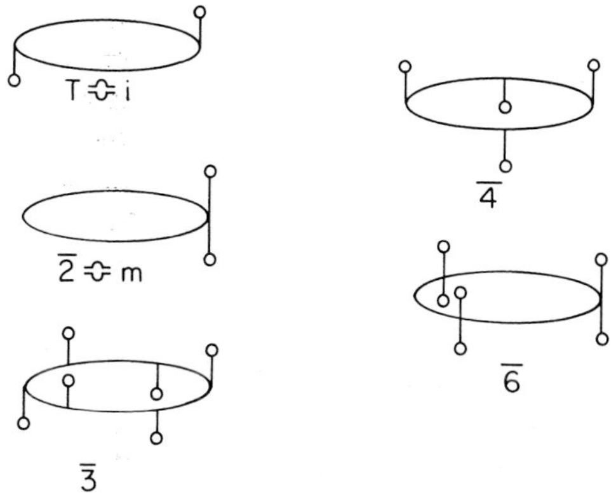

Fig. III.2. Symmetries of the improper axes

addition of inversion as a symmetry element) and are called rotoinversion operations. Figures which portray the symmetries generated by the resulting rotoinversion axes ($\bar{1}$, $\bar{2}$, $\bar{3}$, $\bar{4}$ and $\bar{6}$, for, as will be shown below, rotoinversion (improper) axes are compatible with translational periodicity only if they are of the same order as the allowed proper axes) are shown in Fig. III.2. Note that while the odd improper axes imply a center of symmetry, the even improper axes do not. Considering symmetry elements, the $\bar{1}$ axis is equivalent to an inversion center, the $\bar{2}$ axis to a mirror plane, the $\bar{3}$ axis to a 3-fold proper axis and a center of inversion and the $\bar{6}$ axis to a 3-fold axis and a horizontal mirror plane. Only the improper 4-fold axis is not equivalent to any other element or combination of symmetry elements.

Since the operations of the $\bar{1}$ and $\bar{2}$ axes are operations of inversion and reflection, respectively, it is possible when discussing the symmetry operations of a crystalline solid to refer to the **rotations** (proper and improper) and thereby include in this category inversions and reflections, making it unnecessary to say, "the rotations and the reflections and the inversions" when these symmetry operations are discussed.

III.3 Combination of Rotations and Translations

As was done earlier it is possible to represent combined rotational and translational symmetry operations in a single matrix, for example the C_{6z} operation and translation by **a** in D_{6h} is represented by the 4×4 matrix (Seitz operator)

$$\begin{pmatrix} 1 & \bar{1} & 0 & 1 \\ 1 & 0 & 0 & 0 \\ 0 & 0 & 1 & 0 \\ 0 & 0 & 0 & 1 \end{pmatrix} \begin{pmatrix} x \\ y \\ z \\ 1 \end{pmatrix} = \begin{pmatrix} x - y + 1 \\ x \\ z \\ 1 \end{pmatrix} \tag{III.3}$$

which takes x, y, z into $x - y + 1$, x, z, as shown. This combined operation can be symbolized $C_{6z} \mid \mathbf{a}$ or, a general combined symmetry operation, can be symbolized by $\beta \mid \mathbf{t}$. The β part represents the 3×3 submatrix in the upper

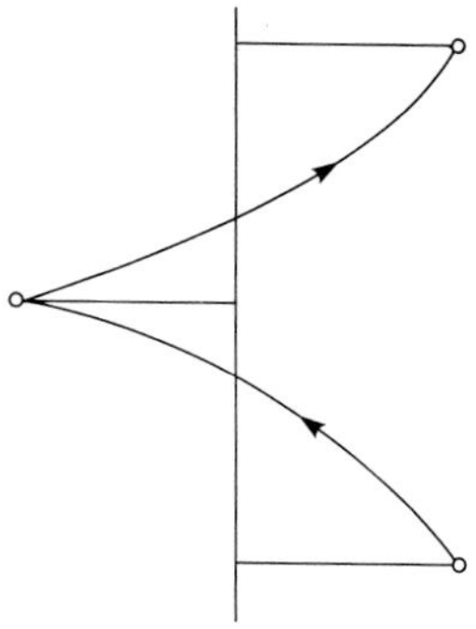

Fig. III.3. The operation of a 2_1 axis

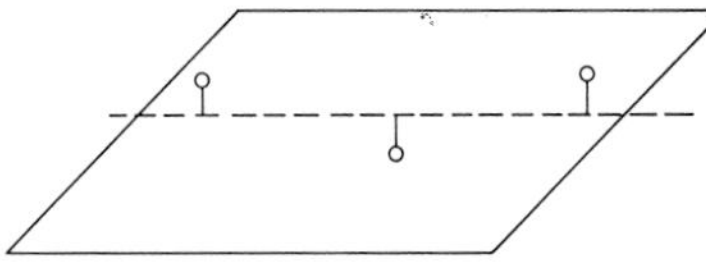

Fig. III.4. The glide symmetry operations

left of the 4×4 matrix i.e., represents the rotational (proper or improper) operator, while **t** represents the 3×1 submatrix on the top of the last column i.e., represents the translational part of the symmetry operation. The last row of the 4×4 matrix is always 0 0 0 1, a device which makes the matrix multiplication work out to appropriately transform the positional parameters, when they are in a 4×1 column matrix with a 1 in the bottom row.

The symbol **t** was intentionally chosen (as opposed to **T**) because in space group symmetries **t** need not be a lattice translation. This occurs in particular when the operation is a screw or glide operation, as discussed below.

III.4 Screw Axes an Glide Planes

First consider how the operations $\beta \mid \mathbf{t}$ combine. Consideration of the multiplication of the 4×4 matrices demonstrates that the β parts combine just as do the 3×3 matrices representing the rotations in the absence of a translational component (Tables III.1 and III.2), Furthermore, the translation part of the operation performed first is rotated by the rotation of that performed second and combined with the translational component of that performed second, i.e.,

$$\beta_2 \mid \mathbf{t}_2 \cdot \beta_1 \mid \mathbf{t}_1 = \beta_2\beta_1 \mid \beta_2\mathbf{t}_1 + \mathbf{t}_2 \,. \tag{III.4}$$

If $\beta \mid \mathbf{t}$ is a combined rotation-translation operation, then operation n times with this operation, where $\beta^n = \varepsilon$, yields

$$\varepsilon \mid \beta^{n-1}\mathbf{t} + \beta^{n-2}\mathbf{t} + \ldots + \mathbf{t} \tag{III.5}$$

which must be a pure translation if $\beta \mid \mathbf{t}$ is to be a possible symmetry operation. For example consider $C_2 \mid \mathbf{c}/2$, where

$$C_2 \mid \mathbf{c}/2 \cdot C_{2z} \mid \mathbf{c}/2 = \varepsilon \mid \mathbf{c} \,. \tag{III.6}$$

Since **c** is a pure translation it follows that $C_{2z} \mid \mathbf{c}/2$ is a possible symmetry operation. This operation is one of a 2_1 screw axis as shown in Fig. III.3. This operation is represented by

$$\begin{pmatrix} \bar{1} & 0 & 0 & 0 \\ 0 & \bar{1} & 0 & 0 \\ 0 & 0 & 1 & \frac{1}{2} \\ 0 & 0 & 0 & 1 \end{pmatrix}$$

which takes x, y, z into $\bar{x}$, $\bar{y}$, $z + \frac{1}{2}$ (the axis has been taken to pass through $x = y = 0$). The product of the matrix with itself,

$$\begin{pmatrix} \bar{1} & 0 & 0 & 0 \\ 0 & \bar{1} & 0 & 0 \\ 0 & 0 & 1 & \frac{1}{2} \\ 0 & 0 & 0 & 1 \end{pmatrix}\begin{pmatrix} \bar{1} & 0 & 0 & 0 \\ 0 & \bar{1} & 0 & 0 \\ 0 & 0 & 1 & \frac{1}{2} \\ 0 & 0 & 0 & 1 \end{pmatrix} = \begin{pmatrix} 1 & 0 & 0 & 0 \\ 0 & 1 & 0 & 0 \\ 0 & 0 & 1 & 1 \\ 0 & 0 & 0 & 1 \end{pmatrix}, \tag{III.7}$$

in agreement with the above, represents **c**, a pure translation. Note that such rotational operations of a space group containing a 2_1 axis do not form a subgroup of the space group since the rotational space group operations are not closed under binary combination.

Applying $\beta^n = \varepsilon \rightarrow \beta^{n-1}\mathbf{t} + \beta^{n-2}\mathbf{t} + \ldots + \mathbf{t} = \mathbf{T}$ to $\beta = C_2$, C_3, C_4 and C_6 (which will be shown to yield all of the possible screw axes) yields the axes summarized in Table III.3. The **rotational symmetry operations** can be defined as those operations for which the β part is not ε, and the remaining operations are the pure translations $\{\varepsilon \mid \mathbf{T}_{mnp}\}$. The rotational symmetry operations in space groups can then be divided into essential and implied operations, where the essential operations are those which contain no pure translational part (no $\mathbf{T}_{mnp}$) and the implied operations are those which result from essential operations by combination with pure translations. It is convenient to know how to locate an arbitrary axis relative to an origin (usually taken as a lattice point). A general screw-rotational operation, $\beta \mid \mathbf{t} + \mathbf{t}'$ has a translation component along the axis about which β operates, (**t**), and a component perpendicular to the axis (**t**′). If **e** is a vector

Table III.3. The Allowed Screw Axes Nm, where N Represents the Rotation and m/N the Fractional Translation Along the Axis

				2_1				
		3_1				3_2		
	4_1			4_2			4_3	
6_1		6_2		6_3		6_4		6_5

from the origin to the axis and perpendicular to the axis, then the total operation can be considered to be performed by:

1. translation by $-\mathbf{e}$.
2. rotation by β about the origin,
3. translation by $\mathbf{e}$,
4. translation by $\mathbf{t}$, i.e.,

$$\begin{aligned}\beta \mid \mathbf{t} + \mathbf{t}' &= \varepsilon \mid \mathbf{e} \cdot \beta \mid 0 \cdot \varepsilon \mid -\mathbf{e} + \varepsilon \mid \mathbf{t} \\ &= \beta \mid \beta(-\mathbf{e}) + \mathbf{e} + \mathbf{t}\,,\end{aligned} \tag{III.8}$$

and

$$\mathbf{t}' = \beta(-\mathbf{e}) + \mathbf{e}\,. \tag{III.9}$$

Thus the existence in $\beta \mid \mathbf{t} + \mathbf{t}'$ of a component of translation perpendicular to the axis implies that the rotation is about an axis which does not pass through the origin. For example

$$\begin{pmatrix} 1 & 0 & 0 & \frac{1}{2} \\ 0 & \bar{1} & 0 & \frac{1}{2} \\ 0 & 0 & \bar{1} & \frac{1}{2} \\ 0 & 0 & 0 & 1 \end{pmatrix}$$

which can be written $C_{2x} \left| \dfrac{\mathbf{a} + \mathbf{b} + \mathbf{c}}{2} \right.$, is an operation of a 2_1 axis along x through a point given by

$$\frac{\mathbf{b} + \mathbf{c}}{2} = C_{2x}(-\mathbf{e}) + \mathbf{e}\,. \tag{III.10}$$

Since $\mathbf{e}$ is perpendicular to the axis, $C_{2x}(-\mathbf{e}) = \mathbf{e}$ and thus $\mathbf{e} = \dfrac{\mathbf{b} + \mathbf{c}}{4}$ i.e. the 2_1 axis for which the above matrix represents an operation passes through the unit cell at the points with $y = z = \frac{1}{4}$.

Another kind of combination of a β operation with a non-lattice translation that occurs as a symmetry operation in space groups is that of reflection combined with translation parallel to the plane of reflection (Fig. III.4). Since reflection is its own inverse the translation $2\mathbf{t}$ parallel to the plane must be a translational symmetry operation. The glide operations that result are labeled according to the direction of the glide, i.e., an **a** glide has translation component parallel to the reflection plane ($\mathbf{a}/2$), an n-glide is a diagonal glide across a face, and a d-glide is diagonal across a centered face.

It is possible that a mirror or glide plane does not pass through the origin of the unit cell, for example while

$$\begin{pmatrix} 1 & 0 & 0 & \frac{1}{2} \\ 0 & 1 & 0 & \frac{1}{2} \\ 0 & 0 & \bar{1} & 0 \\ 0 & 0 & 0 & 1 \end{pmatrix}$$

represents as n-glide perpendicular to **c** with the reflection plane containing the origin,

$$\begin{pmatrix} 1 & 0 & 0 & \frac{1}{2} \\ 0 & 1 & 0 & \frac{1}{2} \\ 0 & 0 & \bar{1} & \frac{1}{2} \\ 0 & 0 & 0 & 1 \end{pmatrix}$$

represents the same operation with the plane of reflection at $z = \frac{1}{4}$. The location of the plane of reflection is determined by the component of the translation perpendicular to the reflection plane. In this case

$$\mathbf{c}/2 = \sigma\,(-\mathbf{e}) + \mathbf{e} \qquad \text{(III.11)}$$

and

$$\mathbf{e} = \tfrac{1}{4}\mathbf{c} \quad \text{(see Fig. III.5).} \qquad \text{(III.12)}$$

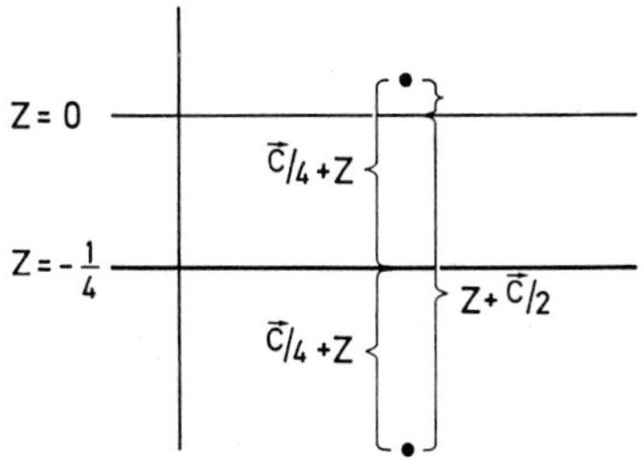

Fig. III.5. Reflection through $z = 0$ followed by translation by $\mathbf{c}/2$ is equivalent to reflection through $\mathbf{c}/4$

III.5 Combination of Symmetry Operations

The rotational symmetry of a lattice that is consistent with a space group with only the rotational symmetry operations of a 2_1 axis is 2, with only those of a 3_1 axis is 3, with those of a 6_3 axis is 6, and so forth. That is, the rotational symmetries of a lattice corresponding to a given space group are the β parts of the $\beta \mid \mathbf{t}$ operations of the space group. To prove this is so, consider a space group in which $\beta \mid \mathbf{t}$ and $\varepsilon \mid \mathbf{T}$ are symmetry operations. The product

$$\beta \mid \mathbf{t} \cdot \varepsilon \mid \mathbf{T} \cdot \beta^{-1} \mid -\beta^{-1}\mathbf{t} = \varepsilon \mid \beta\mathbf{T} \qquad \text{(III.13)}$$

must be a symmetry operation because $\beta^{-1} \mid \beta^{-1}\mathbf{t}$ is the inverse of $\beta \mid \mathbf{t}$ ($\beta^{-1} \mid -\beta^{-1}\mathbf{t} \times \beta \mid \mathbf{t} = \varepsilon \mid 0$) and therefore is a symmetry operation. Thus if $\beta \mid \mathbf{t}$ is a symmetry operation of a solid then β is a rotational symmetry operation of the translations, which is what we set out to prove.

It also follows from this result that if β is an operator for an improper rotational symmetry operation of the space group then the lattice also has the improper rotational symmetry β. Since all lattices are centrosymmetric through lattice points, the lattices consistent with solids with improper axes will have at least the symmetry required by the addition, when it is not implied by the axis, of an inversion center to the improper axis, i.e., $\bar{2}$ in the structure implies 2-fold lattice symmetry, $\bar{3}$ in the structure implies 3-fold lattice symmetry, $\bar{4}$ in the structure implies 4-fold lattice symmetry and $\bar{6}$ in the structure implies 6-fold lattice symmetry. Thus lattices consistent with structures with n-fold rotoinversion axes have n-fold proper axes, and the lattices consistent with n-fold screw axes have n-fold proper axes. It follows that the only allowed improper axes in space groups and the only allowed screw axes are 1-fold, 2-fold, 3-fold, 4-fold and 6-fold, as stated previously.

The use of the 4×4 matrices in the consideration of space group operations and their combinations is illustrated by the following problem, „what are the nature and location of the symmetry operations of the space group Pnma“? The space group symbol Pnma means: a primitive cell (P) with an n-glide perpendicular to **a**, a mirror perpendicular to **b** and an **a** glide perpendicular to **c**. Thus the β (rotational) parts of the symmetry operations are σ_x, σ_y and σ_z from which it follows that the crystal class is D_{2h}. Since it is not clear at the outset where to place the glides and mirror relative to the origin, we take the translational components perpendicular to the reflection planes to be unspecified: $\mathbf{t}_1$, $\mathbf{t}_2$ and $\mathbf{t}_3$. Then the n-glide perpendicular to **a** is represented by

$$\begin{pmatrix} \bar{1} & 0 & 0 & t_1 \\ 0 & 1 & 0 & \frac{1}{2} \\ 0 & 0 & 1 & \frac{1}{2} \\ 0 & 0 & 0 & 1 \end{pmatrix}$$

the mirror perpendicular to **b** by

$$\begin{pmatrix} 1 & 0 & 0 & 0 \\ 0 & \bar{1} & 0 & t_2 \\ 0 & 0 & 1 & 0 \\ 0 & 0 & 0 & 1 \end{pmatrix}$$

and the a-glide perpendicular to **c** by

$$\begin{pmatrix} 1 & 0 & 0 & \frac{1}{2} \\ 0 & 1 & 0 & 0 \\ 0 & 0 & \bar{1} & t_3 \\ 0 & 0 & 0 & 1 \end{pmatrix}.$$

Three two-fold-type (i.e., two-fold or two-fold screw) operations are generated by pairwise combination of these reflection-type operations:

$$\begin{pmatrix} \bar{1} & 0 & 0 & t_1 \\ 0 & 1 & 0 & \frac{1}{2} \\ 0 & 0 & 1 & \frac{1}{2} \\ 0 & 0 & 0 & 1 \end{pmatrix} \begin{pmatrix} 1 & 0 & 0 & 0 \\ 0 & \bar{1} & 0 & t_2 \\ 0 & 0 & 1 & 0 \\ 0 & 0 & 0 & 1 \end{pmatrix} = \begin{pmatrix} \bar{1} & 0 & 0 & t_1 \\ 0 & \bar{1} & 0 & t_2 + \frac{1}{2} \\ 0 & 0 & 1 & \frac{1}{2} \\ 0 & 0 & 0 & 1 \end{pmatrix}, \qquad \text{(III.14)}$$

$$\begin{pmatrix} \bar{1} & 0 & 0 & t_1 \\ 0 & 1 & 0 & \frac{1}{2} \\ 0 & 0 & 1 & \frac{1}{2} \\ 0 & 0 & 0 & 1 \end{pmatrix} \begin{pmatrix} 1 & 0 & 0 & \frac{1}{2} \\ 0 & 1 & 0 & 0 \\ 0 & 0 & \bar{1} & t_3 \\ 0 & 0 & 0 & 1 \end{pmatrix} = \begin{pmatrix} \bar{1} & 0 & 0 & t_1 - \frac{1}{2} \\ 0 & 1 & 0 & \frac{1}{2} \\ 0 & 0 & \bar{1} & t_3 + \frac{1}{2} \\ 0 & 0 & 0 & 1 \end{pmatrix}, \qquad \text{(III.15)}$$

$$\begin{pmatrix} 1 & 0 & 0 & 0 \\ 0 & \bar{1} & 0 & t_2 \\ 0 & 0 & 1 & 0 \\ 0 & 0 & 0 & 1 \end{pmatrix} \begin{pmatrix} 1 & 0 & 0 & \frac{1}{2} \\ 0 & 1 & 0 & 0 \\ 0 & 0 & \bar{1} & t_3 \\ 0 & 0 & 0 & 1 \end{pmatrix} = \begin{pmatrix} 1 & 0 & 0 & \frac{1}{2} \\ 0 & \bar{1} & 0 & t_2 \\ 0 & 0 & \bar{1} & t_3 \\ 0 & 0 & 0 & 1 \end{pmatrix} \qquad \text{(III.16)}$$

Next, the combination of any one of the two-fold-type operations with the perpendicular reflection-type operation will yield an inversion:

$$\begin{pmatrix} \bar{1} & 0 & 0 & t_1 \\ 0 & \bar{1} & 0 & t_2 - \frac{1}{2} \\ 0 & 0 & 1 & \frac{1}{2} \\ 0 & 0 & 0 & 1 \end{pmatrix} \begin{pmatrix} 1 & 0 & 0 & \frac{1}{2} \\ 0 & 1 & 0 & 0 \\ 0 & 0 & \bar{1} & t_3 \\ 0 & 0 & 0 & 1 \end{pmatrix} = \begin{pmatrix} \bar{1} & 0 & 0 & t_1 - \frac{1}{2} \\ 0 & \bar{1} & 0 & t_2 + \frac{1}{2} \\ 0 & 0 & \bar{1} & t_3 + \frac{1}{2} \\ 0 & 0 & 0 & 1 \end{pmatrix}. \qquad \text{(III.17)}$$

The location of this inversion center in the unit cell is fixed by fixing t_1, t_2, t_3. If the conventional choice of origin at the inversion center is made then $t_1 = \frac{1}{2}$, $t_2 = -\frac{1}{2}$ and $t_3 = -\frac{1}{2}$ or, since $+\frac{1}{2}$ and $-\frac{1}{2}$ are equivalent as regards location of symmetry operations we can take $t_1 = t_2 = t_3 = \frac{1}{2}$. Substituting this result into the matrices for the essential symmetry operations yields the following matrices (the corresponding essential symmetry elements are also given):

$$\begin{pmatrix} \bar{1} & 0 & 0 & \frac{1}{2} \\ 0 & 1 & 0 & \frac{1}{2} \\ 0 & 0 & 1 & \frac{1}{2} \\ 0 & 0 & 0 & 1 \end{pmatrix} \longrightarrow \text{an n-glide} \perp \text{to } \mathbf{a} \text{ at } \mathbf{a}/4\,,$$

$$\begin{pmatrix} 1 & 0 & 0 & 0 \\ 0 & \bar{1} & 0 & \frac{1}{2} \\ 0 & 0 & 1 & 0 \\ 0 & 0 & 0 & 1 \end{pmatrix} \longrightarrow \text{a mirror} \perp \text{to } \mathbf{b} \text{ at } \mathbf{b}/4,$$

$$\begin{pmatrix} 1 & 0 & 0 & \frac{1}{2} \\ 0 & 1 & 0 & 0 \\ 0 & 0 & \bar{1} & \frac{1}{2} \\ 0 & 0 & 0 & 1 \end{pmatrix} \longrightarrow \text{an a-glide} \perp \text{to } \mathbf{c} \text{ at } \mathbf{c}/4,$$

$$\begin{pmatrix} \bar{1} & 0 & 0 & \frac{1}{2} \\ 0 & \bar{1} & 0 & 0 \\ 0 & 0 & 1 & \frac{1}{2} \\ 0 & 0 & 0 & 1 \end{pmatrix} \longrightarrow \text{a } 2_1 \text{ axis} \parallel \text{to } \mathbf{c} \text{ at } x = \frac{1}{4}, \qquad y = 0$$

$$\begin{pmatrix} \bar{1} & 0 & 0 & 0 \\ 0 & 1 & 0 & \frac{1}{2} \\ 0 & 0 & \bar{1} & 0 \\ 0 & 0 & 0 & 1 \end{pmatrix} \longrightarrow \text{a } 2_1 \text{ axis} \parallel \text{to } \mathbf{b} \text{ through } x = z = 0\,,$$

$$\begin{pmatrix} 1 & 0 & 0 & \frac{1}{2} \\ 0 & \bar{1} & 0 & \frac{1}{2} \\ 0 & 0 & \bar{1} & \frac{1}{2} \\ 0 & 0 & 0 & 1 \end{pmatrix} \longrightarrow \text{a } 2_1 \text{ axis} \parallel \text{ to } \mathbf{a} \text{ through } y = z = \frac{1}{4},$$

$$\begin{pmatrix} \bar{1} & 0 & 0 & 0 \\ 0 & \bar{1} & 0 & 0 \\ 0 & 0 & \bar{1} & 0 \\ 0 & 0 & 0 & 1 \end{pmatrix} \longrightarrow \text{an inversion center at the origin.}$$

Other, implied symmetry operations in the unit cell can be generated by the combination of those given above with pure translations, for example the 2_1 axis parallel to **c** at $x = \frac{1}{4}$, $y = 0$ combines with translation by **b** to yield a 2_1 axis parallel to **c** at $x = \frac{1}{4}$, $y = \frac{1}{2}$, etc.

In the chapters that follow the group properties of space groups become important. It is therefore important to bear in mind that the symmetry operations of crystalline solids combine associatively, are closed under "multiplication", and include inverses for all operations and the identity operation i.e., the set $\{\beta_i \,|\, \mathbf{t}\}$ is a group. The set of all β_i's, $\{\beta\}$, is a point group called the crystal class. To each of the 32 crystal classes there corresponds a number of space groups that differ from each other in the sense that the essential axes may or may not be screw axes and the essential reflection planes may or may not be glide planes. For example the crystal class $C_{2h}(\varepsilon, C_2, \sigma_y$ and i) is consistent with a monoclinic lattice. To this crystal class belong the primitive space groups: P2/m, $P2_1/m$, P2/c, and $P2_1/c$. Among these only P2/m does not include screw axes or glide planes among its essential operations. Such a space group is called **symmorphic**. The other, nonsymmorphic, space groups can be considered to result from the systematic, allowed replacement of the pure rotational operations

Table III.4

Class C_{2h}	C_{2y}	i	σ_y	ε
P2/m	$C_{2y}\|0$	$i\|0$	$\sigma_y\|0$	$\varepsilon\|0$
$P2_1/m$	$C_{2y}\|\frac{\mathbf{b}}{2}$	$i\|0$	$\sigma_y\|\frac{\mathbf{b}}{2}$	$\varepsilon\|0$
P2/c	$C_{2y}\|\frac{\mathbf{c}}{2}$	$i\|0$	$\sigma_y\|\frac{\mathbf{c}}{2}$	$\varepsilon\|0$
$P2_1/c$	$C_{2y}\|\frac{\mathbf{b}+\mathbf{c}}{2}$	$i\|0$	$\sigma_y\|\frac{\mathbf{b}+\mathbf{c}}{2}$	$\varepsilon\|0$

of P2/m by rotation-translation operations. The essential operations of the four groups with the origin taken to be at the inversion center are given in Table III.4.

III.6 Problems

1. Find the symmetry operation of O_h that results from the combination of a 3-fold rotation about x + y + z and a 4-fold rotation about y.
2. Prove that a 5_1 axis cannot occur in a crystalline solid.
3. Show explicitly that a 4_2 axis is consistent with a cubic or tetragonal lattice.
4. Describe the essential symmetry elements (type and location) of:
   a. Pbam.
   b. P23 (a two-fold and a three-fold which are in directions $\alpha = \cos^{-1}(1/\sqrt{3})$) — what is the lattice type?
   c. $P6_3/m$.
5. Find the location of the two-fold screw axes in Immm. Is this space symmorphic? Explain.
6. Provide a list of positions equivalent by symmetry to x, y, z in $P2_1/c$ with the origin at the center of symmetry.
7. Show that $P2_1/m$ is a subgroup of P2/m.
8. Find the crystal class for each of the following space groups.
   a. Aba2.
   b. $P4_2/m$.
   c. R32.
   d. Pa3.
9. Find the space group symmetries that result from the addition of inversion symmetry to $P2_12_12$.

Chapter IV

A Brief Sampling of Some Inorganic Structure Types

IV.1 Introduction

Much of the discussion of solid state chemistry takes for granted a familiarity with the structures of a number of key structures and structure types. It is therefore necessary to become familiar with a variety of crystal structures and their descriptions and interrelationships. Two detailed and, for most purposes, complete reference works with which students of solid state chemistry should become familiar are Structural Inorganic Chemistry by A. F. Wells [25] and the Handbook of Lattice Spacings and Structures of Metals, Vol. 2 by W. B. Pearson [26]. Only a few selected structures are discussed here with the view of providing an initial introduction to a vast subject.

Many insights and points of discussion turn on the understanding that chemists have of the features of structures, and these depend to some extent upon how the structures are viewed. For example, the NaCl-type structure, a very simple and common structure type described in Table IV.1 and shown schematically in Fig. 1.1, can be viewed as cubic-close-packed anions with cations in the octahedral holes, as MY_6 octahedra sharing edges with resultant moderately short M—M distances, as AbCaBc . . . (upper case nonmetals, lower case metals) sphere packing (see Fig. IV.1), or as ccp metal with nonmetals in the octahedral interstices as discussed below. Since all of the above descriptions refer to the same structure they do have common features. However, the different descriptions place emphases on different structural aspects and thus represent differing points of view.

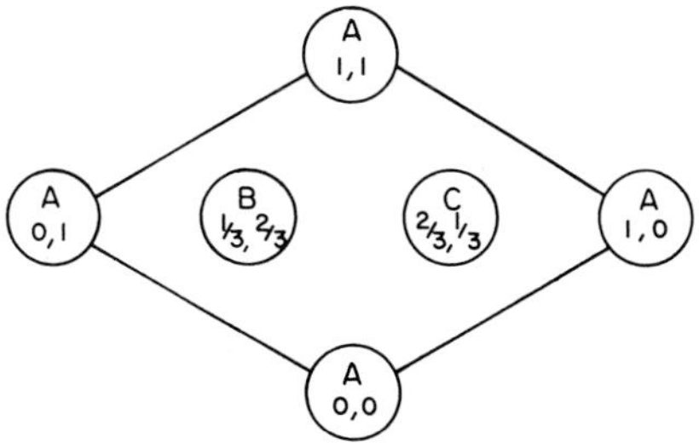

Fig. IV.1

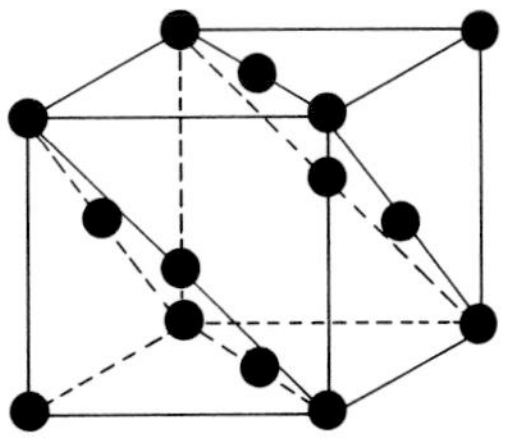

Fig. IV.2

Fig. IV.1. Hexagonal stacking positions projected onto the a-b plane A = 0, 0; B = $\frac{1}{3}, \frac{2}{3}$; C = $\frac{2}{3}, \frac{1}{3}$.

Fig. IV.2. Face centered cubic structure from cubic closest packing (ABCABC . . .).

Table IV.1

hcp	Ba(II), α-Be, Cd, α-Co, Dy, Er, Fe(II), Gd, α-Hf, Ho, Li(L.T.), Lu, Mg, Na(L.T.), Os, Re, Ru, Sc, β-Sr, α-Tb, Te, α-Ti, α-Tl, Tm, Y, Zn, α-Zr
ccp	Ac, Ag, Al, Am(H.T.), Au, α-Ca, γ-Ce, α-Ce(L.T.), β-Co, Cu, γ-Fe, Ir, β-La(H.T.), La(H.P.), γ-Mn(H.T.), Nd(H.P.), Ni, Pb, Pd, Pr(H.P.), Pt, δ-Pu(H.T.), Rh, α-Sr, α-Th, Yb

One approach to the systematization of structure that has been part of almost every discussion of inorganic crystalline solids from the beginning is the approach based upon the structures of densely packed spheres. When hard spheres of equal size are packed so as to minimize the amount of void space they form planar sheets with hexagonal symmetry in which each sphere is surrounded by six others in a regular hexagon with the distance from sphere center to sphere center being in each case twice the sphere radius (2r). In the closest packed arrangements, the planar sheets are stacked in such a fashion that the spheres of adjacent sheets form regular tetrahedra, i.e., the projected positions of the sphere centers from the second layer fall in the centers of the equilateral triangles with edge length 2r formed by triples of spheres in the first layer.

There are two sets of centers upon which the projected positions may fall, and these are labeled B and C if the positions in the basal sheet are labeled A (see Fig. IV.1). There are two simple ordered examples of packing of this type, called hexagonal closest packing (hcp), in which the layers stack ABABAB . . ., and cubic closest packed (ABCABCABC . . .) which have numerous analogues in the crystal structures of the elemental metals (see Table IV.1). It is important to emphasize that the hard sphere scheme is one for the description of structure, not for the rationalization of structure. There has been and continues to be a great deal of discussion of the electronic structures and their impact upon the structures adopted by the elementary metals, and the notion that these structures form because the atoms are basically spherically symmetrical does not appear to be taken seriously by anyone. One immediate problem with such a simple view would be the obvious problem that if structures such as hcp and ccp formed only because the atoms packed as spheres, then there would be no reason for the structures to form as ordered ABABAB . . . or ABCABCABC . . . analogues of sphere packing, but rather disordered structures (e.g.. ABACBCABCBAC . . .) would be predominant.

The names hexagonal closest packed and cubic closest packed derive from the lattice symmetries of the structures. The ABABAB . . . stacking provides a hexagonal lattice with sphere centers at $\frac{1}{3}, \frac{2}{3}, \frac{1}{4}$ and $\frac{2}{3}, \frac{1}{3}, \frac{3}{4}$. The ABCABCABC . . . stacking, on the other hand, gives rise to a face-centered cubic lattice (see Fig. IV.2). Note that ABCABCABC . . . is exactly the same as the stacking that leads from plane hexagonal lattices to a rhombohedral lattice and that this rhombohedral lattice is also face centered cubic when α is the interedge angle of a regular tetrahedron (i.e., 60°). The extent to which space is filled by hard spheres packed in a ccp arrangement can be determined from the volume of the fcc unit cell ($= A^3 = (2\sqrt{2}\,r)^3$) and the volume occupied by the corresponding number of spheres ($= 4 \cdot \frac{4}{3}\pi r^3$) i.e., fraction of space filled by ccp hard spheres $= (16\pi r^3/3)/(r^3/2^{9/2}) = 0.74$. It is left as an exercise to prove that the same result is obtained for hcp hard spheres.

The void space, which is 26% of the total, can be described as consisting of tetrahedral (e.g., centered at $\frac{1}{4}, \frac{1}{4}, \frac{1}{4}$ in the cubic cell) and octahedral (e.g., centered at $\frac{1}{2}, \frac{1}{2}, \frac{1}{2}$ in the cubic cell) "interstices" (see Fig. IV.2). By the symmetry of the cubic cell there are eight equivalent tetrahedral interstices (centered at $\frac{1}{4}, \frac{1}{4}, \frac{1}{4}; \frac{1}{4}, \frac{1}{4}, \frac{3}{4}; \frac{1}{4}, \frac{3}{4}, \frac{1}{4}; \frac{3}{4}, \frac{1}{4}, \frac{1}{4}; \frac{3}{4}, \frac{3}{4}, \frac{3}{4}; \frac{3}{4}, \frac{3}{4}, \frac{1}{4}; \frac{3}{4}, \frac{1}{4}, \frac{3}{4}$ and $\frac{1}{4}, \frac{3}{4}, \frac{3}{4}$) and four equivalent octahedral interstices (centered at $\frac{1}{2}, \frac{1}{2}, \frac{1}{2}; \frac{1}{2}, 0, 0; 0, \frac{1}{2}, 0$ and $0, 0, \frac{1}{2}$) per unit cell. Some frequently occurring structures can be described as analogues of those that would be obtained by the filling of these interstices with hard spheres with sufficiently small radii that the small spheres would fit into the interstices. The sizes of spheres which would just fit into the interstitial positions can be calculated from purely geometric considerations and yield the radius ratios which have been frequently cited. Straightforward calculation yields $(\sqrt{3} - \sqrt{2})/\sqrt{2} = 0.2247$ for the radius ratio for filling of the tetrahedral voids, and 0.4142 for the octahedral voids. Again with the caveat that this description is by analogy, Table IV.2 is presented to provide a few structural descriptions of some crystal structures that will be discussed at various points later in the book.

A small selection of particularly simple crystal structure types important in Solid State Chemistry is presented in Table IV.3. This table lists in the first column the structure type and its space group symbol from which, by the methods of Chap. III, the equivalent positions within a unit cell can be worked out in detail. The second column lists the special positions on which the metal (M) and nonmetal (Y) atoms are located. The third and fourth columns give the nearest neighbor like and unlike atom coordination polyhedra, and the last three columns list the shortest distances in terms of the lattice parameters. These shortest distances have traditionally been compared with expected distances based upon tabulated values of radii. The radii used differ from case to case since a number of consistent sets of radii are available. In the discussions of the various structures presented below the following comparisons are made:

Table IV.2

Structure type	Nonmetals	Metals
NaCl	ccp	all octahedral interstices
CaF_2	all tetrahedral interstices	ccp
ZnS	half of the tetrahedral interstices	ccp
NiAs	hcp	all trigonal antiprismatic interstices
CdI_2	hcp	$\frac{1}{2}$ of the octahedral sites
$CdCl_2$	ccp	$\frac{1}{2}$ of the octahedral sites
Sc_2S_3	ccp	$\frac{2}{3}$ of the octahedral sites

Table IV.3

Structure type Space Group		x, y, z	Y coord.	M coord.	d_{MY}	d_{MM}	d_{YY}
NaCl	M	0, 0, 0	octahedral	ccp	$\frac{a}{2}$ (6x)	$\frac{a}{\sqrt{2}}$ (12x)	$\frac{a}{\sqrt{2}}$ (12x)
Fd3m	Y	0, 0, $\frac{1}{2}$	ccp	octahedral			
WC	M	0, 0, 0	trig. prism	hex. bipyr.	$\sqrt{\frac{a^2}{3}+\frac{c^2}{4}}$	a(6x)	a(6x)
P$\bar{6}$2m	Y	$\frac{1}{3}, \frac{2}{3}, \frac{1}{2}$	hex. bipyr.	trig. prism		c(2x)	c(2x)
CsCl	M	0, 0, 0	cubic	oct.	$\frac{a\sqrt{3}}{2}$ (8x)	a(6x)	a(6x)
Pm3m	Y	$\frac{1}{2}, \frac{1}{2}, \frac{1}{2}$	oct.	cubic			
NiAs	M	0, 0, 0	trig. antiprism	hex. bipyr.	$\sqrt{\frac{a^2}{3}+\frac{c^2}{16}}$	a(6x)	a(6x)
$P6_3/mmc$	Y	$\frac{1}{3}, \frac{2}{3}, \frac{1}{4}$	hcp	trig. prism		c(2x)	$\sqrt{\frac{a^2}{3}+\frac{c^2}{4}}$ (6x)
ZnS	M	0, 0, 0	tetrahedral	ccp	$\frac{a\sqrt{3}}{4}$ (4x)	$\frac{a}{\sqrt{2}}$ (12x)	$\frac{a}{\sqrt{2}}$ (12x)
F$\bar{4}$3m	Y	$\frac{1}{4}, \frac{1}{4}, \frac{1}{4}$	ccp	tetrahedral			
CaF_2	M	0, 0, 0	cubic	ccp	$\frac{a\sqrt{3}}{4}$, M – Y(8x)	$\frac{a}{\sqrt{2}}$ (12x)	$\frac{a}{2}$ (6x)

Fm3m	Y	$\frac{1}{4}, \frac{1}{4}, \frac{1}{4}$	octahedral	tetrahedral	Y—M (4x)	$\frac{a}{\sqrt{2}}$ (12x)
ReO_3	M	0, 0, 0	octahedral	—	$\frac{a}{2}$, M—Y (6x)	
Pm3m	Y	$\frac{1}{2}$, 0, 0	tet. prism	linear	Y—M (2x)	$\frac{a}{\sqrt{2}}$ (8x)

1) the metal-nonmetal distance (d_{MY}) with the sum of the Slater [4] radii,
2) the nonmetal-nonmetal distance (d_{YY}) with twice the van der Waals radii, and
3) the metal-metal distance (d_{MM}) with a distance between metals characteristic of the elemental solid metal.

These comparisons result in a number of conclusions, the most important of which is that the radii provide at best only a very rough guide to the interpretation of solids. Other conclusions, which are necessarily rather uncertain because of lack of precision in comparisons based upon radii, are interspersed in the following discussions of structures.

IV.2 NaCl-Type

The first entry in the table, the NaCl-type, is a very common structure type which plays an important role in later chapters. This structure type is known for oxides, carbides, nitrides, sulfides, phosphides, selenides, arsenides, tellurides, antimonides, polonides and bismuthides of the transition metals, rare earths, and actinides, as well, of course, for some alkali halides and alkaline earth chalcides. Many of the compounds exhibit interesting and useful properties such as high temperature stability, metallic conductivity and extensive defect chemistry.

Two rather typical NaCl-type materials which will be used as examples in later chapters for discussions of electronic structure, defect chemistry and defect ordering are TiO and ScS. One characteristic of many inorganic solids that has been used extensively in consideration of structure is the fact that in different compounds with the same M and Y elements the metal to nonmetal distances d_{MY} are roughly the same from case to case leading to tabulated sets of radii which differ from author to author, but which, when summed, yield typical d_{MY} values characteristic of the solid state. For example, the Slater [4] radii of Sc and S are 160 and 100 pm, respectively (each uncertain by about 5 pm, according to Slater), and these values yield a predicted ScS lattice parameter of 520 ± 7 pm. The observed value is 516.5 pm [19]. The corresponding prediction for TiO is 400 ± 7 pm and the observed value is 418 pm [19], however it is important to note that stoichiometric (O/Ti = 1.00) TiO occurs with approximately 15% of the O and Ti sublattices vacant, perhaps accounting for the rather substantial discrepancy.

Also of note is the fact that $d_{MM} = d_{YY} = \sqrt{2}\, d_{MY}$ for this structure type. To a first approximation d_{YY} is limited from below by the sum of the van der Waals radii, and relatively short d_{MM} values are associated with metal-metal interactions in the conducting transition-metal compounds. Pauling [10] gives 140 pm for the van der Waals radius of O (yielding $a \geqq 280\sqrt{2} = 396$ pm, which fits for TiO), and 185 pm for that of S (yielding $d_{MS} \geqq 261.6$ pm which, when compared with $d_{ScS} = 258.3$ pm, implies either that the suggested van der Waals radius is somewhat too large, or that the stability of ScS is adversely affected by S-S repulsions).

The d_{MM} values in TiO (296 pm) and ScS (365 pm), when compared with the d_{MM} values in the metals (about 286 pm and 326 pm, respectively [26]), suggest the possibility of weak metal-metal interactions. In the case of Sc the difference of nearly 40 pm is rather large, and hence the nature and extent of the interactions leading to metallic conductivity is not immediately obvious. Note that since $d_{MM} = d_{YY}$ is required in this structure type there is a competition between small values for d_{MM} (presumably correlated with stabilizing metal-metal interactions) and large values for d_{YY} (presumably correlated with small values for the destabilizing nonmetal-nonmetal interactions). The metal-metal interactions in ScS are discussed in detail in Chap. X.

One important aspect of the chemistry of TiO and ScS that should be mentioned is that both compounds are known to exhibit wide ranges (10—20% of the sites vacant is a "wide" range) of homogeneity. TiO is known to exist for $0.64 \leqq O/Ti \leqq 1.25$ while ScS is known to exist for $0.9 \leqq S/Sc \leqq 1.24$ (the upper limit is probably even higher). Furthermore these vacancies are observed to order when samples are cooled slowly. The ordered structures are discussed thoroughly in Chaps. VIII and IX.

In summary, the NaCl-type structure is one that is adopted by a variety of metal-nonmetal binaries, including refractory, conducting, transition-metal compunds which exhibit wide ranges of homogeneity and concomitant order-disorder transitions. The variety of physical and chemical properties which are exhibited by these compounds provides a fruitful area for the exploration of a developing understanding of the interrelationship between structure, stability and electronic structure.

IV.3 NbO-Structure

The structure of NbO (space group Pm3m) [26] is related to the NaCl-type structure in an interesting way, namely the Nb atoms are located in the center of each face ($\frac{1}{2}$, 0, $\frac{1}{2}$; $\frac{1}{2}$, $\frac{1}{2}$, 0 and 0, $\frac{1}{2}$, $\frac{1}{2}$) but not at the origin, and O atoms are located in the middle of each edge ($\frac{1}{2}$, 0, 0; 0, $\frac{1}{2}$, 0; 0, 0, $\frac{1}{2}$) but not in the center of the cell. In other words, the NbO-type structure could be viewed as an ordered defect NaCl-type structure, and as resulting from a hypothetical order-disorder process (i.e., $\frac{3}{4}$ of the metal and nonmetal positions of NaCl-type occupied at random transforming to the NbO-type structure via differentiation of the occupancies of the origin position (lower occupancy) and the face centering positions (higher occupancy) for the metal atoms, and differentiation of the cell center position (lower occupancy) and the edge-center positions (higher occupancy) for the nonmetal atoms). However, such a consideration is purely theoretical.

The symmetry of the coordination of both Nb by O and O by Nb in NbO is square planar with $d_{NbO} = 210.5$ pm (compared with the sum of the Slater radii, 205 ± 7 pm), and $d_{NbNb} = d_{OO} = 298$ pm (8x) compared with $d_{NbNb} = 286$ pm in metallic Nb (26) and twice the van der Waals radius of O which equals 280 pm. The symmetry of the Nb coordination by Nb is tetragonal.

IV.4 WC-Type

The WC-type structure (Fig. IV.3) provides an interesting contrast to the NaCl-type. One of the descriptions possible for the NaCl structure is that of ccp of spheres, but with adjacent layers of spheres (along the cubic [1, 1, 1] directions)

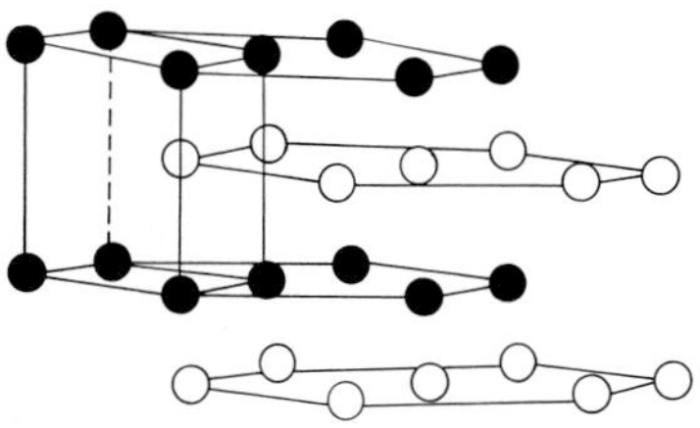

Fig. IV.3

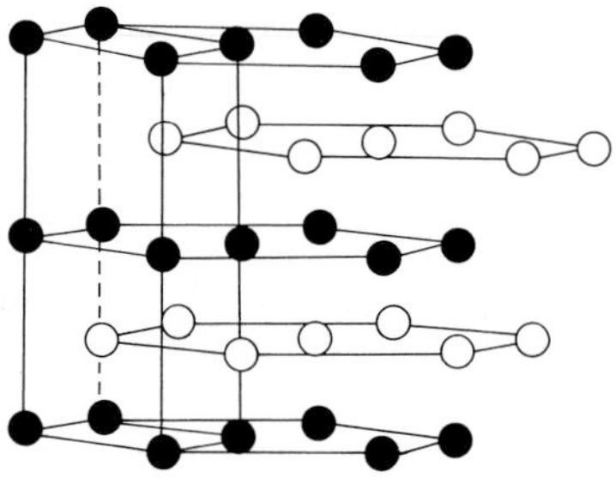

Fig. IV.4

Fig. IV.3. WC-type structure emphasizing the AbAb stacking of alternate hexagonal layers of metals and nonmetals.

Fig. IV.4. NiAs-type structure emphasizing the AbAc ... stacking of alternate hexagonal layers of metals and nonmetals.

having different chemical identities (i.e., M and Y). The WC-type structure is then a hcp (ABAB ... layering) form of this structural configuration. It is not possible without making some assumption with regard to the c/a ratio to compare distances in the WC and NaCl type structures. If WC did arise from the packing of equal size M and Y spheres then $d_{MY} = d_{YY}$ and $\sqrt{a^2/3 + c^2/4} = a$, or c/a $= \sqrt{8/3} = 1.633$. However, c/a has been found to be 1.00 ± 0.02 for three known WC-type materials, WC, ZrS and HfS [19], indicating that close packing of equal size spheres is far from a reasonable geometric interpretation of the WC structure in these transition-metal compounds.

Taking c/a = 1.00 for the purposes of calculation yields, for WC-type, $d_{YY} = d_{MM} = \sqrt{12/7}\, d_{MX}$ and since for NaCl-type $d_{YY} = d_{MM} = \sqrt{2}\, d_{MX}$, the d_{MM} and d_{YY} values are 8% longer in the NaCl structure type than in the WC structure type for the same d_{MX}. Thus if there is a like atom repulsive interaction at distances characteristic of the WC structure with a given d_{MY}, then those repulsions can be substantially relieved without altering d_{MY} by transition to the NaCl-type structure. It is perhaps for this reason that there are many more NaCl-type solids than there are WC-type solids.

ZrS is typical among the WC-type compounds. It is metallic and refractory, and it shares much in the way of appearance (metallic) and properties with ScS. Zirconium deficient $Zr_{1-x}S$ transforms to an ordered defect rock-salt structure. Nearly stoichiometric sulfur deficient ZrS_{1-x} (which occurs for small values of x) has the WC-type structure. For WC-type ZrS (19) $d_{ZrS} = 263$ pm (the sum of the Slater radii is 255 ± 7 pm) and the d_{SS} values are 343 pm (6x) and 345 pm (2x) (the S—S van der Waals distance is 2×185 pm = 370 pm, again suggesting that 185 pm is too large for a van der Waals radius for sulfur in solids of this type), and $d_{Zr-Zr} = 343$ pm (6x) and 345 pm (2x) to be compared with 315 pm in the metal. As with the Sc—Sc distance in ScS, this metal—metal distance is about 40 pm longer in ZrS than in metal. A more detailed investigation of the interactions leading to the metallic character is left to Chap. X.

IV.5 NiAs-Type

In a sense the NiAs-type structure (Fig. IV.4) strikes a compromise between the NaCl-type and WC-type structures. Whereas, NaCl-type can be described as AbCaBc ... stacking of hexagonal layers (along the cubic [1, 1, 1] directions), and WC-type can be described as AbAb ... stacking of hexagonal layers along the hexagonal c direction, the NiAs-type structure can be described as aBaCaBaC ... stacking (the a layers are the metal layers, the B and C layers are the nonmetal layers) of hexagonal layers along the hexagonal c direction.

The compromise provides for a closer approach of metal to metal (along c (2x)) than does the NaCl-type structure and a greater interplanar separation of nonmetals than does the WC-type structure. The structure has at times been described as hcp nonmetals with metals in the octahedral interstices. If the nonmetals were hcp then $\sqrt{\frac{a^2}{3} + \frac{c^2}{4}}$ (d_{YY}, of one kind, see Table IV.3) would equal a (d_{YY} of the other kind), and $c/a = \sqrt{\frac{8}{3}} = 1.633$, and the coordination of metal by nonmetal would be octahedral. For real compounds with $c/a \neq 1.633$ the actual coordination is a trigonal antiprism, a trigonal distortion of the octahedron. For the many transition metal compounds (sulfides, phosphides, selenides, arsenides, etc.) which form with this structure type, c/a ranges between 1.27 for PtBi to 1.96 for VP [26].

The compounds TiS and VS form with the NiAs-type structure, although stoichiometric VS forms with this structure type only at high temperature, and deforms to the MnP-type structure at lower temperatures. This transition is discussed in detail in Chap. VII. TiS is somewhat anomalous because it has a very large (second only to VP) c/a ratio (1.93) [19]. This fact provides a basis for testing the importance of the short M—M distance along c in stabilizing the NiAs structure through M—M interactions, and these electronic interactions in TiS are discussed in Chap. X. In TiS (19) $d_{TiS} = 265$ pm (the sum of the Slater radii is 240 ± 7 pm), the short $d_{TiTi} = c/2 = 319$ pm compared to 286 pm in Ti metal [26], and $d_{YY} = a = 330$ pm compared with 370 pm for twice the van der Waals radius. Once again it appears that the van der Waals radius for sulfur is too large and that the metal-metal distance is significantly larger than in the elemental metal, but certainly not too large for some Ti—Ti interaction. For VS, $d_{VS} = 255$ pm (the sum of the Slater radii is 235 pm), d_{VV} is 291 pm (d_{VV} in elemental V(s) is 262 pm [26], and $d_{SS} = 333.3$ pm, again to be compared with 370 pm, twice the van der Waals radius for sulfur.

The comparison given in all three cases (NaCl, WC and NiAs) between the observed distances and those obtained from Slater radii, elemental metal and van der Waals radii are clearly less than satisfactory as a basis for a detailed understanding of the structures and the bonding features stabilizing them. The comparions are provided to indicate some first-order generalizations concerning the transition metal compounds (d_{MY} is roughly the same from compound with the same M and Y in "contact", hence radii such as Slater's covalent radii can be tabulated and used to obtain an estimate of d_{MY}, d_{MM} in transitional-metal compounds is frequently not substantially greater than that found in the elemental metal, suggesting possible metal-metal interactions and the nonmetal-nonmetal distances are to a rough approximation bounded from below by a sum of van der Waals radii).

IV.6 CsCl-Type

The CsCl structure type (Fig. IV.5) is unknown for the type of compounds discussed above, i.e., carbides, nitrides, oxides, phosphides, sulfides, etc., but is a very common structure type for intermetallic compounds [26]. The nominally 1:1 Cu—Zn alloy with a homogeneity range, for example, is one of the many known CsCl-type intermetallics. This compound is called β′ bass, and at high temperatures this solid

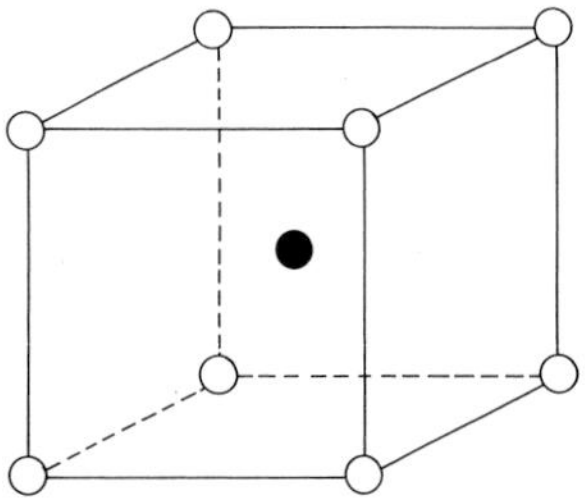

Fig. IV.5

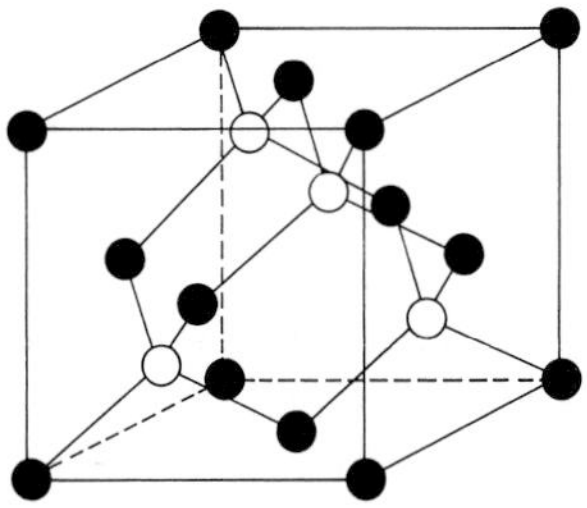

Fig. IV.6

Fig. IV.5. CsCl-type structure.
Fig. IV.6. ZnS-type structure emphasizing the tetrahedral coordination and relationship to the diamond-type structure.

transforms continuously to β brass, which has the bcc structure with each lattice site of a bcc lattice occupied at random by Cu or Zn. Such a solid with random occupation is bcc only on the average, since of course, each site is occupied either by Cu or Zn (or a vacancy or impurity, but these minor effects will be neglected for the immediate purposes of discussion). However, because the methods of determining longer range crystal structure (discussed in Chap. IX) depend upon the scattering of waves (X-rays, electrons or neutrons) which interact with a large number of atoms on a large number of lattice sites, it follows that the structure determined is an average structure, e.g., for stoichiometric β-brass the determined structure is one with "half a Cu and half a Zn" at each bcc lattice site (in this case neutron diffraction is far more useful than X-ray diffraction because Cu and Zn differ by only one electron or about 3% in X-ray scattering power). The transition from CsCl-type to bcc is a classic example of an order-disorder transition, and is discussed further in Chaps. VII and IX.

IV.7 ZnS-Type

The cubic ZnS (zinc blende)-type structure (Fig. IV.6) is related to the diamond structure in the sense that if Zn and S were equivalent the structure would be of the diamond type. Unlike the structure types discussed above, which are well known for their tendency to occur as metallic transition metal compounds (NaCl, WC and NiAs-type) and intermetallic compounds (NaCl and CsCl types), as well as insulating 1-1 and 2-2 compounds (e.g., NaCl, CsCl, CaO, etc.), the cubic ZnS-type structure is best known for its tendency to form for 3–5 semiconductors (AlAs, AlP, AlSb, BAs, GaAs, InAs, BN, InP, etc.), although it is also known for some 2—6 semiconductors (BeS, BeSe, CdS, CdSe, HgS, HgSe, ZnO, ZnS, etc.). Lacking, however, are examples of conducting transition-metal compounds such as form with the NiAs, WC and NaCl-type structures.

In ZnS the Zn—S distance is 235 pm (sum of the Slater radii is 235 pm) and the Zn—Zn and S—S distances are 383 pm, substantially longer, respectively, than the distance found in metallic Zn(s) (266 pm) and longer than twice the sulfur van der Waals radius. One possible description of the cubic ZnS structure is ccp metal (or nonmetal) spheres with nonmetal (or metal) in $\frac{1}{2}$ the tetrahedral holes, however in order for this description to be realistic the ratio of the radii of the spheres must be such that the smaller sphere just fits into the tetrahedral interstitial position. The ratio of larger sphere radius to smaller sphere radius must then be greater than $\sqrt{2}/(\sqrt{3} - \sqrt{2}) = 4.45$. Taking the tabulated Zn and S radii with the greatest difference, namely the ionic radii of Zn^{2+} and S^{-2} (74 and 184 pm, respectively), yields a radius ratio of 2.49, far below the value required for ccp S^{-2} with Zn^{+2} in $\frac{1}{2}$ the tetrahedral holes. Note also that the sum of the ionic radii, 258 pm, is not in good agreement with the observed d_{ZnS}.

IV.8 CaF_2-Type

The CaF_2 structure can be thought of as the filled ZnS structure, i.e., the Ca atoms are in the ccp positions with all of the resulting tetrahedral sites filled with F atoms. Some examples of solids with the fluorite (or anti-fluorite, i.e., M_2Y as opposed to MY_2) structure are Be_2C, a number of rare-earth dihydrides, Mg_2Si, a number of rare-earth, actinide and other dicarbides, dinitrides, dioxides, including VC_2, VN_2 and VO_2, Li_2O and related compounds. Cubic eight coordination is a feature of this solid. For VO_2, as an example, $d_{VO} = 237$ pm [26] compared with the sum of the Slater radii 235 pm, $d_{VV} = 387$ pm compared with $d_{VV} = 275$ pm, the shortest distance in the elemental metal, and $d_{OO} = 274$ pm, compared with twice the van der Waals radius, 280 pm. A variety of novel solid-state materials, e.g., LiZnP, Zn_3P_2 and $TlCu_2Se_2$ ($ThCr_2Si_2$-type) have structures related to CaF_2-type.

IV.9 The ReO_3 Structure

The ReO_3 structure type (Fig. IV.7) is not widely adopted, Cu_3N and ReO_3 are known in this structure type. It is, however, a convenient starting point for the discussion of the sheared oxide structures that have been a major object of study (principally by electron microscopy) and discussion in solid state chemistry, and for the perovskite structure which has also played an important role in solid state science.

The ReO_3-type structure can be viewed as ReO_6 octahedral sharing all corners.

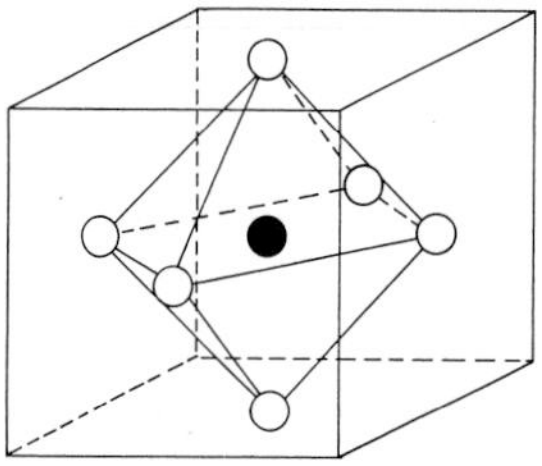

Fig. IV.7. ReO_3 structure with M at $\frac{1}{2}, \frac{1}{2}, \frac{1}{2}$; Y at $\frac{1}{2}, 0, \frac{1}{2}$; $0, \frac{1}{2}, \frac{1}{2}$; $\frac{1}{2}, \frac{1}{2}, 0$ emphasizing octahedral coordination of M by Y.

There are no short M—M interactions, the shortest distance between metals being the M—Y—M distance which is equal to $|\mathbf{a}|$ (= 375 pm in ReO_3) [26]. In ReO_3 d_{ReO} = 187.5 pm, which compares with 195 pm, the sum of the Slater radii, and d_{OO} = 265 pm (twice the van der Waals radius of O is 280 pm).

IV.10 The Layered Dichalcides

The layered dichalcides, i.e., principally the disulfides and diselenides of the group IV, V and VI transition metals, have been extensively studied because of their interesting intercalation properties, catalytic activity, superconductivity, charge density wave behavior and photoelectrical stability. The solids consist of stacks of hexagonal layers occurring in the sequence (Y means nonmetal hexagonal layer, M means metal hexagonal layer): ...YMYYMYYMY.... The resultant structures may have rhombohedral or trigonal (principal axis of the structure is three-fold with, respectively, a rhombohedral or hexagonal lattice type) or hexagonal (principal axis of the structure is six-fold) symmetry. These are labeled, respectively, R, T and H.

All of the special positions at which atoms are located in the hexagonal cells of the R, T and H structures lie in the 110 plane (see Fig. IV.1), and thus the structures can be completely represented by a view of this plane. Some examples are shown in Fig. IV.8. The first of these, 1T, is the structure type of TiS_2, for example. The structure type is called CdI_2-type. Its space group symmetry is $P\bar{3}m1$. The

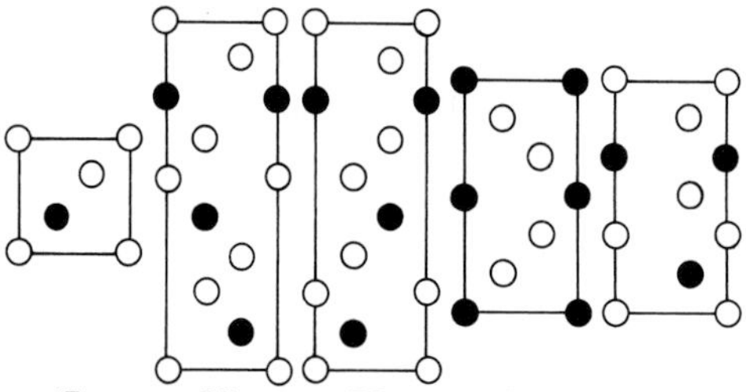

Fig. IV.8. Slice through the hexagonal cells of a variety of layered dichalcide structure types. Dark circles indicate metal-atom positions, light circles indicate nonmetal-atom positions. Note that the structures can be described as YMY hexagonal layer sandwiches held together by van der Waals interactions.

stacking sequence is AbCAbC.... When compared with the NiAs-type structure, ...AbCbAbCb..., it appears that the CdI_2-type structure is metal deficient NiAs-type (alternate metal layers missing) or, alternatively that the NiAs-type structure is an odered solid solution resulting from intercalation of M in MY_2.

Since TiS_2 has the CdI_2-type structure and TiS has the NiAs-type structure the thought arises that perhaps a complete range of solid solution might occur with $Ti_{1-x}S$ ($x \leqq 1$) occurring with each alternating metal layer completely filled and with 2x vacant metal positions distributed at random throughout the sites in the partially filled layers. It has been shown, however, that what does occur instead is a succession of Ti_mS_n phases with different stacking sequences and population wave occupancy. This behavior will be discussed in Chap. IX. Another interesting possibility exists, namely that for some NiAs-type compound with a deficiency of metal atoms, that the vacancies will be for some thermodynamic states (taken here to be fixed by temperature, pressure and composition) randomly distributed over all metal positions (symmetry characteristic of NiAs-type with less than one M at each M site) and for some other thermodynamic states, e.g., lower temperatures, the vacancies will be randomly distributed within alternate metal containing planes, the intervening planes being completely occupied (symmetry characteristic of CdI_2-type with fractional occupancy taking the place of the unoccupied positions). The transition from the described NiAs-type to CdI_2-type symmetries is another example of an order-disorder transition, and this described transition is known for chromium deficient $Cr_{1-x}S$ at high temperatures (see Sect. IX.5).

The second structure type shown for the layered dichalcides in Fig. IV.8, one of the 3R structures, has an ACaBAbCBcA... stacking sequence which can be compared with the AbCaBc sequence of the NaCl-type structure. It follows that this structure can be viewed as an ordered vacancy NaCl-type structure in a fashion exactly analogous to that just discussed for CdI_2-type relative to NiAs-type. The structure type that results for MY_2 in this case is the $CdCl_2$ type (space group $R\bar{3}m$), which is the structure type of a number of dihalides.

The ordering of vacancies in alternate metal-containing layers also occurs in $Sc_{1-x}S$, which exhibits the NaCl-type to $CdCl_2$-type analogue of the NiAs-type to CdI_2-type phase transition. At high temperatures the NaCl-type structure is observed for scandium deficient monosulfide (defects randomly distributed) and at low temperatures the vacancies order such that alternate metal containing planes (along one of the body diagonals) are equivalent ($R\bar{3}m$ symmetry). This order-disorder transition, which apparently occurs as a second-order phase transition, will be discussed further in Chap. IX.

The other variants of the layered dichalcides shown in Fig. IV.8 show some of the variety that results from variation in stacking order. The Y—M—Y layer units can be seen to be of two types, namely those with two dissimilar Y layers and those with two similar Y layers. The former is the case when the Y layers adjacent to a given metal layer have different projected positions on the basal plane, whereas, the latter occurs when the projected positions are the same. In the former case the metal coordination is trigonal antiprismatic (as in NiAs-type) and in the latter the metal coordination is trigonal prismatic (as in WC-type). The electronic structures in these two cases are substantially different, as will be discussed in Chap. X.

IV.11 Perovskite

The $CaTiO_3$-type structure (space group Pm3m, Ca at 0, 0, 0; Ti at $\frac{1}{2}, \frac{1}{2}, \frac{1}{2}$ and O at $0, \frac{1}{2}, \frac{1}{2}$) is closely related to the ReO_3 structure, i.e., it arises from ReO_3 by filling the cube origin position that is empty in ReO_3. This structure type is found for a number of ternary nitrides and carbides such as $AlTi_3C$ which is perhaps better expressed in this context as $AlCTi_3$ [26], since in this compound there is a network of Ti octahedra (d_{TiTi} = 294 pm) with C in the octahedral holes and Al in 12 coordination (Fig. IV.9). The perovskite structure has played an important role in solid state science. It is, for example, the structure type of cubic sodium tungsten bronze ($NaWO_3$). Furthermore, the perovskite structure has been observed to undergo numerous distortions, for example, $BaTiO_3$ [27] appears rhombohedral (a = 400.1 pm, α = 89,95°) to X-rays (although physical measurements indicate a lower symmetry) at low temperatures (below 183 K), is orthorhombic (space group Amm^2, a = 399.0, b = 566.9 and c = 568.2 pm) for $183 < T < 263$ K, tetragonal (P4mm, a = 399.4 and c = 403.8 pm) for $263 < T < 397$ K and cubic perovskite for higher temperatures. Finally $BaTiO_3$ transforms to a hexagonal form at very high temperatures.

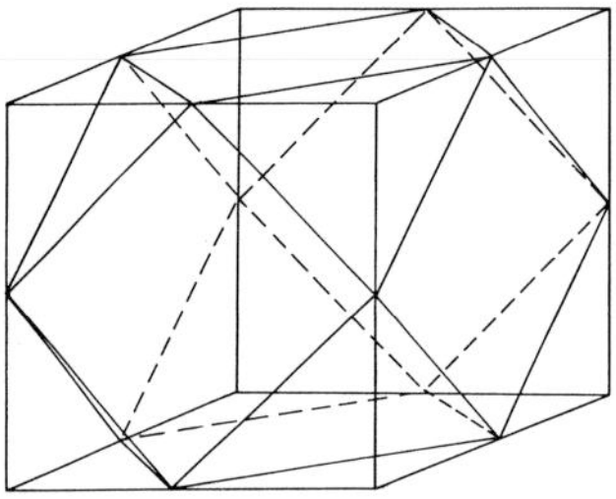

Fig. IV.9

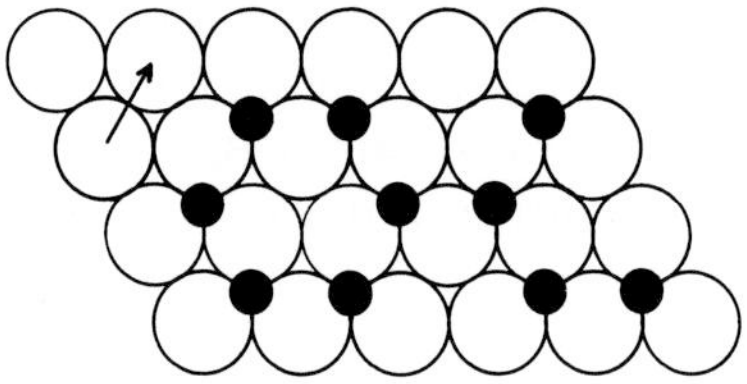

Fig. IV.10

Fig. IV.9. Central atom coordination polyhedron for the perovskite structure.

Fig. IV.10. Al_2O_3-type structure: Idealized M atom positions (dark circles) on a close packed layer of Y atom positions. Metal atom positions in consecutive layers are shifted by the vector shown in the upper left.

IV.12 α-Al_2O_3

The α-Al_2O_3, or corundum, structure (space group R$\bar{3}$c) is found for a number of sesquioxides (Ti, V, Cr, Fe, Rh and Ga), as well as some rare earth sulfides in one of their crystal modifications. An idealized structure can be described as oxygen in hcp positions with aluminium ordered in $\frac{2}{3}$ of the octahedral holes (Fig. IV.10). The location of the void positions (in terms of projection onto the basal plane) is shifted by the distance and direction of one of the interoxygen vectors consistently from metal layer to metal layer. The result, in so far as the oxygen coordination is concerned, is four coordination at $\frac{2}{3}$ of the corners of a trigonal prism, the vacant corners being diagonally opposed across a rectangular face of the prism. This coordination is distorted tetrahedral. The true corundum structure is obtained from the idealized structure by movement of the metal atoms in the z direction, the two Al's sharing an edge of the trigonal prism move farther apart and the two diagonally opposed on a rectangular face of the prism move closer together. This alteration is in the direction of a more nearly regular tetrahedral coordination of oxygen.

The discription given above for corundum emphasizes its similarly to NiAs (hcp nonmetal atoms with metal atoms in the trigonal antiprismatic holes). No oxide with the NiAs structure is known and trigonal prismatic coordination of oxygen is either unknown or extremely rare. One could imagine filling the appropriate $\frac{1}{3}$ of the octahedral holes in Ti_2O_3 with the α-Al_2O_3-type structure to obtain NiAs-type TiO. However, this does not occur, instead TiO forms with the NaCl-type structure as discussed previously. The nature of oxygen is such that it tends to form solids in which the oxygen coordination is octahedral or tetrahedral rather than trigonal prismatic.

IV.13 Rutile (TiO_2)

The rutile structure (space group $P4_2/mmm$) is observed for a number of dioxides and fluorides. This structure presents a case of triangular planar coordination of the nonmetal and octahedral coordination of the metal atoms. The octahedra form chains sharing edges along the c axis direction, and the chains are linked by corner sharing. This structure is shown in Fig. IV.11.

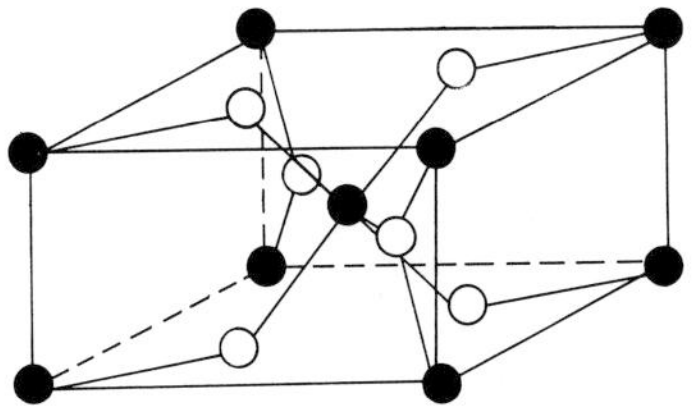

Fig. IV.11

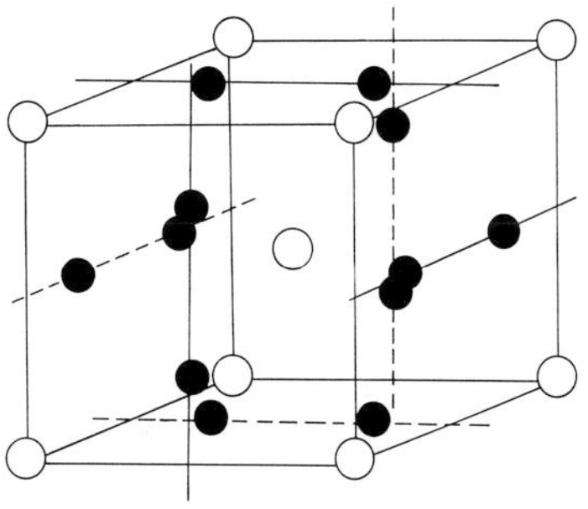

Fig. IV.12

Fig. IV.11. Rutile structure.

Fig. IV.12. W_3O (A15) structure type emphasizing chains of metal atoms in the x, y and z directions.

IV.14 The W_3O or A15 Structure

The A15's, as the compounds with this structure type (Fig. IV.12) are called (see Pearson [26]), crystallize in a cubic cell with fixed positions in Pm3m. This structure type is found most frequently for intermetallic compounds such as Nb_3Al and Nb_3Ge, which are much studied because of their relative high superconducting tranion temperatures. The structure forms with one metal (e.g., Al or Ge) at the corners and the center of a cube and the other metals forming chains with $d_{MM} = a/2$ which run through the centers of the faces parallel to *a*, *b*, *c* in such a fashion that no two such chains intersect. For Nb_3Al the Nb-Nb distance in the chains is $a/2 = 259$ pm (26) (compare with 286 pm in elemental Nb(s) (26)), with next nearest Nb neighbors (8x) at a distance $\sqrt{3/8}\,a = 318$ pm, and Al neighbors (4x) at 290 pm (the sum of the Slater radii is 270 ± 7 pm). As regards the Al coordination, there are twelve nearest Nb neighbors at 290 pm. The coordination polyhedron is describable in terms of parallel equilateral triangles centered upon a body diagonal of the cubic unit cell, two smaller triangles at $\pm\, a\sqrt{3/4}$ along the diagonal and two larger triangles at $\pm\, a/\sqrt{12}$ (Figure IV.13).

IV.15 Metal-Rich Structures

Introduction. There has been an increasing awareness in recent years of the extent to which a variety of chemical and physical properties depend upon the nature and extent of like-atom bonding in chemical compounds. The possibility of relatively weak metal-metal interactions in intrinsically metal-nonmetal compounds was raised in the discussion of transition-metal compounds with the NaCl-type, NiAs-type, WC-type, and related structures in this chapter. The compounds chosen for discussion in this section, as is true for the A15-type, and other, intermetallic structures, provide examples of compounds in which the metal-metal interactions are clearly of substantial importance in stabilizing the solids, i.e., compounds for which

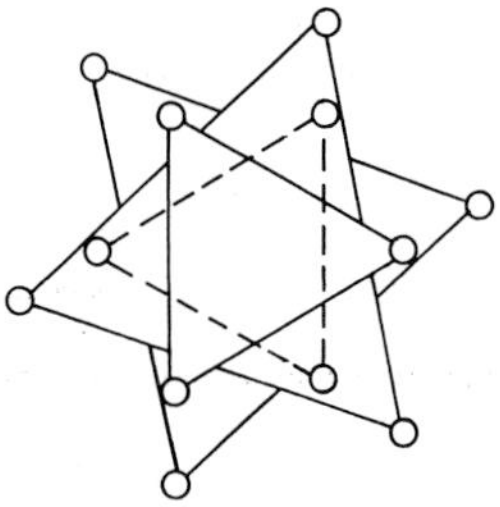

Fig. IV.13

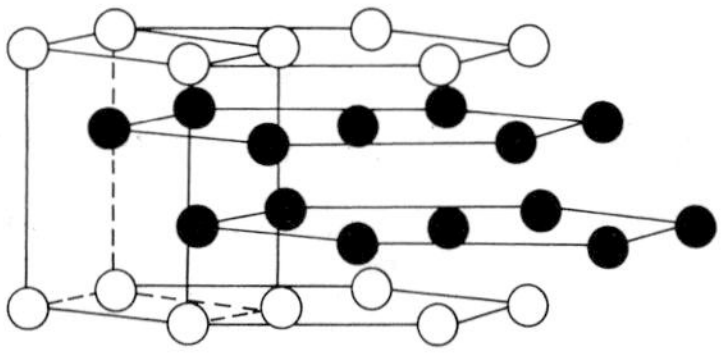

Fig. IV.14

Fig. IV.13. Coordination of Y in M_3Y with the A15 structure type.
Fig. IV.14. Ti_2O structure (anti-CdI_2-type cf. Fig. IV.8).

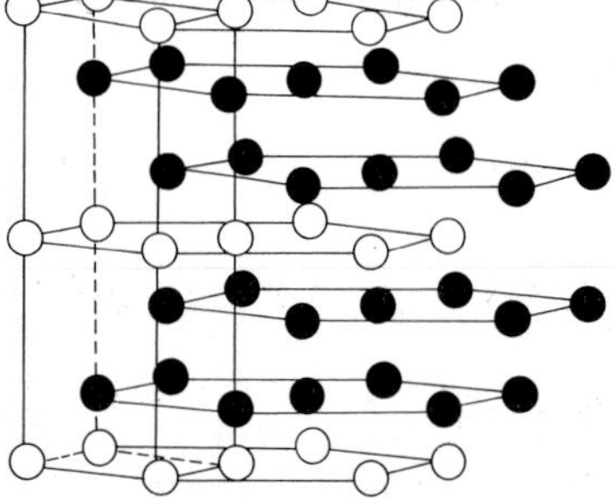

Fig. IV.15. Hf_2S structure (anti2 H—NbS_2-type cf. Fig. IV.8).

no other explanation than one which invokes numerous metal-metal interactions of a strength comparable to that in the elemental metal, is reasonable. In a very real sense the compounds discussed here are "modified metals" where the modification is accomplished through interaction with small amounts (relative to saturated compounds) of characteristic nonmetals such as O, S, N and Cl.

Two simple examples of structures in this category are Ti_2O [28] and Hf_2S [29]. Both structures exhibit MMYMMYMMY. . .layering of hexagonal layers (Fig. IV.1). In the case of Ti_2O the layering is abCabC. . . type (anti-CdI_2-type), while in the case of Hf_2S the layering if AbcAcb. . . type (anti-NbS_2(2H) type). In Ti_2O the nearest neighbor coordination polyhedron of both O and Ti are distorted octahedra, O to six Ti in a trigonal antiprism and Ti to 3O and 3Ti (Fig. IV.14), and in Hf_2S (Fig. IV.15) nearest neighbor coordination of S is trigonal prismatic and that of Hf is a distorted octahedron, again to 3 metals and 3 nonmetals. In Ti_2O the short Ti—Ti distance is 286 pm (3x) (to be compared with d_{TiTi} in Ti(s) of 286 pm [26]) and in Hf_2S the comparable distance is 306 pm (3x) (to be compared with d_{HfHf} in Hf(s) of 312 pm [26]). Note that these short metal-metal distances imply substantial metal-metal interactions, and thus that the relationships between structure type and antitype mentioned above are somewhat misleading since the nearest neighbor nonmetal-nonmetal distances in CdI_2 and NbS_2 are certainly van der Waals distances, while the nearest neighbor metal-metal distances in Ti_2O and Hf_2S are associated with substantial interactions. There is a surprising similarity between the Hf_2S and Ti_2O structures and those of ScCl [30]

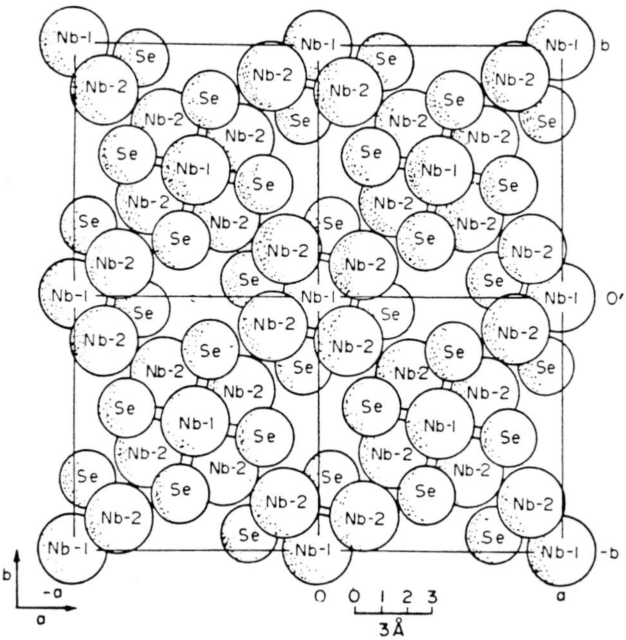

Fig. IV.16. The Ti_5Te_4 structure type shown for Nb_5Se_4 using covalent radii to emphasize space filling and to rationalize both Nb—Nb and Nb—Se distances.

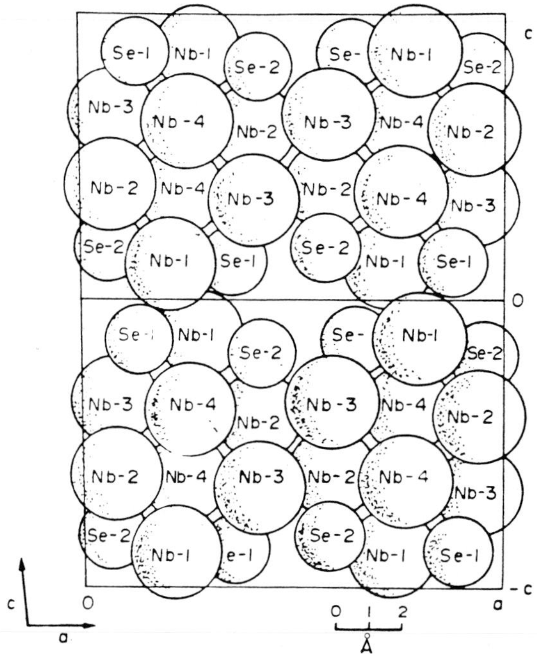

Fig. IV.17. The Nb_2 structure, drawn using covalent radii.

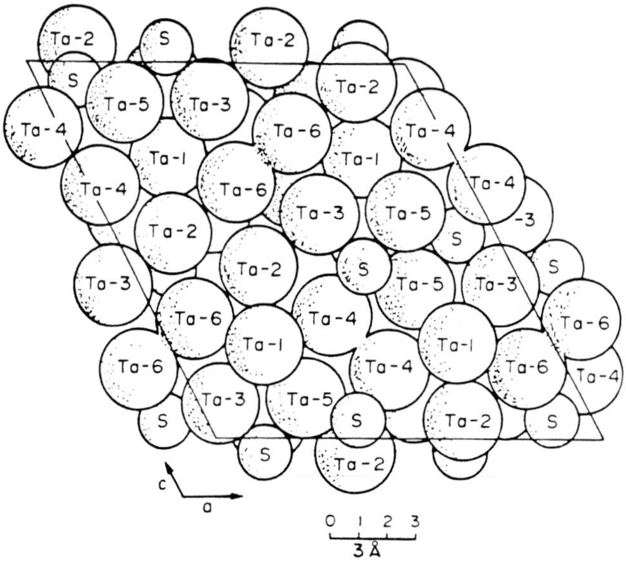

Fig. IV.18. The Ta_6S structure, drawn using covalent radii.

and ZrCl [31] in which the stacking occurs as MMYYMMYY... with abCAbc-ABcaBC...stacking in ZrCl and abCBcaBAbcAC...stacking in ScCl.

A totally different type of structure with substantial metal-metal interactions is the Ti_5Te_4-type of structure shown in Fig. IV.16 for Nb_5Se_4 [32]. In this structure there are columns of body-centered cubes of Nb atoms with Se atoms on the vertical faces, or, alternatively, chains in the c direction of Nb atom octahedra sharing corners and capping atoms with Se atoms on each triangular face. These chains of Nb_6Se_8 units are then linked by interchain interactions. It is interesting to compare the Nb_2Se structure of Fig. IV.17 with that of Nb_5Se_4. In this structure the tendency to fill in the Nb_5Se_4 structure in the direction of bcc Nb is

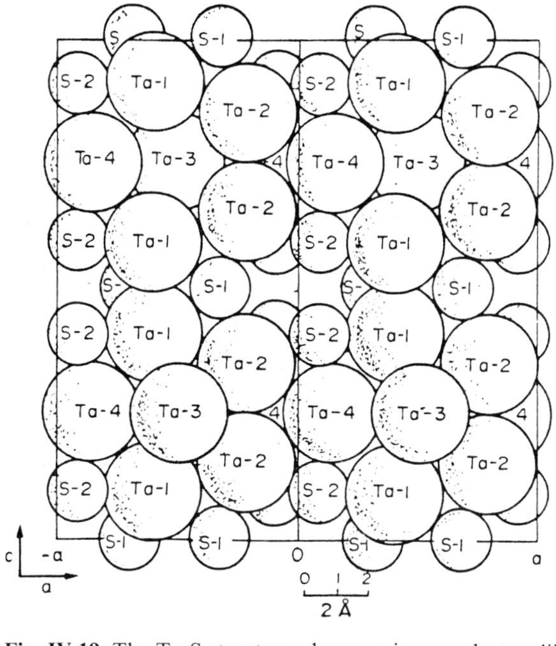

Fig. IV.19. The Ta_2S structure, drawn using covalent radii.

clear, and for Nb_2Se it is possible to suggest that this compound is modified Nb, with Se substitution on Nb sites in the bcc structure resulting in a partial disruption of the structure.

The metal-rich sulfides of Ta are strikingly different from these selenides of Nb. Two Ta-rich sulfides of Ta are known, Ta_6S and Ta_2S (Figs. IV.18 and IV.19). Both structures contain columns of face sharing pentagonal antiprisms of Ta, with sulfur playing a bridging role between the columns. The structure is once again indicative of substantial metal-metal interactions, as is typical of compounds of this type. The metal-rich compounds have been extensively reviewed in recent years, and the interested reader will find a sampling of revealing review articles referenced in Sect. X.3. An examination of these reviews will demonstrate that the few compounds mentioned here are merely an arbitrary small sampling of a rich and varied structural chemistry.

IV.16 Oxide Structures Based on ReO_3

Studies of pure and mixed transition-metal oxides by X-ray diffraction and electron microscopy have shown that there occur a number of reduced (relative to MO_3) oxide structures based upon the ReO_3 structure, but with the introduction of structural modifications to accomodate the stoichiometry change. Two features of the structural modifications make them unusually important features of structural inorganic chemistry. The first is that they introduce the possibility of an enormous variety of structures based upon octahedral coordination of the metal atom and diverse metal-to-nonmetal ratios (M_8O_{23}, M_9O_{26}, $M_{10}O_{29}$, $M_{11}O_{32}$, $M_{12}O_{35}$, $M_{14}O_{41}$, etc.). The second is that the modifications such as lead to the regular stoichiometric structures are observed by electron microscopy to occur also in a disordered way in what are apparently metal-oxide solid solution regions. In the nonstoichiometric octahedral oxides, at least, there is a marked tendency for oxygen vacancies to partially order so as to yield distinct structural modifications, called crystallographic shears, but without the formation of a new long-range three-dimensional periodicity. An example is discussed in Sect. IX.9.

One example of an ordered shear structure, that of Mo_8O_{23} [33] is provided here to illustrate this important structural phenomenon. Figure IV.20a represents the ReO_3-type structure (compare with Fig. IV.7), i.e., Mo_6 octahedra sharing corners. The squares with crosses represent chains of octahedra sharing corners running in a direction perpendicular to the figure.

Figure IV.20b shows the Mo_8O_{23} structure which can be seen to be made of ReO_3-like chains of octahedra sharing corners along the direction perpendicular to the plane of the figure, but with some ordered edge sharing within the plane accompanying the reduction in O/M from 3 to $\frac{20}{8}$. This feature of edge sharing

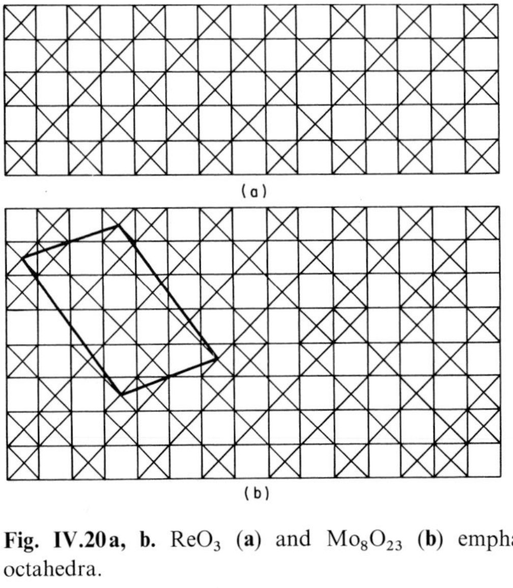

Fig. IV.20a, b. ReO_3 (**a**) and Mo_8O_{23} (**b**) emphasizing edge and corner sharing MO_6 octahedra.

of MO_6 octahedra has been found to be a prominent feature in numerous transition-metal oxides which can be viewed as being based upon the ReO_3-type structure.

IV.17 Problems

1. Na_2Te crystallizes in the CaF_2-type structure with a = 733 pm. What are the Na and Te nearest neighbor coordinations? What is d_{Na+Te}?
2. High temperature NiSe crystallizes in the NiAs-type structure with a = 366 and c = 535 pm. What are the Ni and Se coordinations? What is the shortest d_{Ni-Ni}? What is d_{Ni-Se}?
3. α-Mo_3C_2 is reported to have the NaCl-type structure, yet NaCl is a 1-1 type strcuture. How is the stoichiometry reconciled with the structure type?
4. W_3O crystallizes in the A15 structure type with a = 504 pm. What are d_{W-W} and d_{W-O}?
5. Cu_3N crystallizes in the ReO_3-type structure with a = 381 pm. What is the N coordination? What is d_{Cu-N}?
6. Cu_3Au crystallizes in a cubic structure with Au at 0, 0, 0 and Cu at 0, $\frac{1}{2}$, $\frac{1}{2}$; $\frac{1}{2}$, 0, $\frac{1}{2}$; $\frac{1}{2}$, $\frac{1}{2}$, 0. Sketch the structure. What is its space group?
7. Ni_2In crystallizes with space group $P6_3/mmc$ with Ni(1) in 0, 0, 0; 0, 0, $\frac{1}{2}$; Ni(2) in $\frac{1}{3}$, $\frac{2}{3}$, $\frac{3}{4}$; $\frac{2}{3}$, $\frac{1}{3}$, $\frac{1}{4}$ and In in $\frac{1}{3}$, $\frac{2}{3}$, $\frac{1}{4}$; $\frac{2}{3}$, $\frac{1}{3}$, $\frac{3}{4}$. What is the relationship of this structure to that of NiAs? Sketch the structure.
8. CoSn crystallizes with space group P6/mmm with Sn(1) in 0, 0, 0; Sn(2) in $\frac{1}{3}$, $\frac{2}{3}$, $\frac{1}{2}$; $\frac{2}{3}$, $\frac{1}{3}$, $\frac{1}{2}$ and Co in $\frac{1}{2}$, 0, 0; 0, $\frac{1}{2}$, 0; $\frac{1}{2}$, $\frac{1}{2}$, 0. Sketch the structure.
9. AlB_2 crystallizes with space group P6/mmm with Al in 0, 0, 0 and B in $\frac{1}{3}$, $\frac{2}{3}$, $\frac{1}{2}$; $\frac{2}{3}$, $\frac{1}{3}$, $\frac{1}{2}$. Sketch the structure. For AlB_2, a = 301 and c = 326 pm. What is d_{Al-B}?
10. CrS crystallizes with space group C2/c with a = 383, b = 591, c = 609 pm and $\beta = 101° 36'$. The Cr atoms are in the positions 0, 0, 0; 0, 0, $\frac{1}{2}$ and the sulfur atoms in the positions 0, y, $\frac{1}{4}$; 0, $\bar{y}$, $\frac{3}{3}$ with y = 0.32. Sketch the structure. Describe the Cr coordination.

Chapter V

Thermodynamics

V.1 Introduction

Thermodynamics plays a special role in solid-state chemistry because of the importance of heterogeneous equilibria in the synthesis of many solids of physical and chemical interest. The understanding of such equilibria is made especially important by the occurrence of nonstoichiometry in solids. The nonstoichiometry of solids ranges from very narrow ("line compounds" with negligible ranges of chemical content but usually with important effects on properties such as conductivity) to wide, e.g., 10% or more variation in Y/M. Both cases are considered in this chapter. The understanding of the heterogeneous behavior of systems can be based upon phase diagrams or upon the mathematical relationships of thermodynamics, and is greatly assisted by insight into the interrelationship between these two aspects of the expression of thermodynamic laws. This chapter has as its purpose the development of the general, basic features of these laws as they apply to heterogeneous equilibria.

V.2 Basic Concepts and Assumptions

Thermodynamics, as it is generally considered, consists of, as well as the laws of thermodynamics, a number of concepts and models concerning the macroscopic behavior of matter. Macroscopic means that the concepts and laws of thermodynamics can be developed without reference to the atomic or particulate nature of matter. Thermodynamics can also be developed from a statistical analysis of microscopic properties, but this analysis is not necessary for this discussion and is therefore not developed here. One important concept is that of a reversible process, that is a hypothetical process which occurs through a succession of macroscopic states of rest. For macroscopic states of rest there are well defined macroscopic state variables (P, V, T, X_i, G, S, U, etc.) which are interrelated by equations of state. The consideration of reversible processes, macroscopic states of rest and equations of state do not in any way depend upon the first and second laws of thermodynamics but are independent concepts upon which the development of classical, i.e., reversible, thermodynamics depends.

Of considerable importance in thermodynamics is the assumed behavior of the equations of state, i.e., at the outset it is assumed that the relationships among state variables are well behaved so that the functions are taken to be continuous with continuous partial derivatives except at isolated points (e.g., phase transitions).

Also of the utmost importance is an assumption concerning the numbers of independent variables upon which the consideration of state functions is based, namely that the chemical composition variables can be identified and counted (there are taken to be c of them for a given system), and that in addition to these c composition variables, two additional macroscopic variables are required to fix the state of a system in the absence of long-range (electric, magnetic, gravitational) fields. Much of the basis for the understanding of heterogeneous equilibria, which in turn provides a basis for understanding solid state reactions, nonstoichiometry and phase transitions, all highly important in solid-state chemistry, is the phase rule, a rule which is frequently not developed in a fashion which demonstrates its fundamental nature. Therefore, much of the thermodynamics that follows is directed toward the development of a high level of understanding of the phase rule and of the number of component variables, c.

From the laws of thermodynamic the definitions of three macroscopic properties (state functions) namely U, the internal energy, S, the entropy, and T, the absolute temperature follow directly. Some other macroscopic properties defined without recourse to the laws of thermodynamics are P, the pressure, V, the volume, and N, the number of moles of chemical component, of which there are c independent quantities. In the absence of long-range fields (henceforth understood), c + 2 of these macroscopic properties fix the state of a macroscopic system at rest.

V.3 The Gibbsian Equation for dU

From the above we can take U to be a function of c + 2 variables, e.g., of S, V, N_i with i = 1, 2, ... , c, and can assume that U = U(S, V, N_i) is a well behaved function. Accordingly we write

$$dU = \left(\frac{\partial U}{\partial S}\right)_{V,N_i} dS + \left(\frac{\partial U}{\partial V}\right)_{S,N_i} dV + \sum_{i=1}^{c} \left(\frac{\partial U}{\partial N_i}\right)_{S,V,N_j} dN_i \tag{V.1}$$

where dN_i is the number of moles of ith component exchanged with the surroundings. Making use of the laws of thermodynamics it is a simple matter to show that

$$\left(\frac{\partial U}{\partial S}\right)_{V,N_i} = T\,, \tag{V.2}$$

and

$$\left(\frac{\partial U}{\partial V}\right)_{S,N_i} = -P\,, \tag{V.3}$$

so that

$$dU = T\,dS - P\,dV + \sum_{i=1}^{c} \mu_i\,dN_i \tag{V.4}$$

where, by definition,

$$\mu_i = \left(\frac{\partial U}{\partial N_i}\right)_{S,\,V,\,N_j} \tag{V.5}$$

An important feature of the above is the explicit recognition that the number of independent variables (c + 2) is a basis of, and not a consequence of, this "Gibbsian" Eq. (V.4). This assumption, together with the assumption that the implicit equation of state ($U = U(S, V, N_i)$) is mathematically well behaved (continuous with continuous partial derivatives), and with the definition of a reversible process (so that P, S, μ_i and T remain defined during the process to which dU refers) are all important to the validity of this Gibbsian equation for dU.

V.4 The Phase Rule

The recognition that c + 2 variables fix the state of a system at rest permits the derivation of the phase rule, showing that this rule is a basis of, rather than a result of, the laws of thermodynamics. This derivation is remarkably straightforward, namely we can fix the state of a system at rest as completely as possible by specifying the independent intensive variables, the number of which is called f, and the quantity of each phase, the number of which is called p. Hence, f + p variables fix the state of the system and f + p = c + 2, the phase rule.

V.5 Condition for Heterogeneous Equilibrium

The Gibbsian equation leads directly to the condition for heterogeneous and homogeneous equilibrium. For the heterogeneous case consider phases α, β, γ, ... coexisting in a closed system at contant T and P. For the system as a whole, because it is closed, dU = T dS — P dV for reversible changes. For each phase

$$dU^\alpha = T\, dS^\alpha + P\, dV^\alpha + \sum^{i} \mu_i^\alpha\, dN^\alpha,\ \alpha = 1\ 2\ 3 \ldots p$$

and

$$\sum dU^\alpha = T \sum dS^\alpha - P \sum dV^\alpha + \sum_{\alpha=1}^{p} \sum_{i=1}^{c} \mu_i^\alpha\, dN_i^\alpha \tag{V.6}$$

or

$$dU = T\, dS - P\, dV + \sum_{\alpha=1}^{p} \sum_{i=1}^{c} \mu_i^\alpha\, dN_i^\alpha, \tag{V.7}$$

which by comparison with dU = T dS — PV yields,

$$\sum_{\alpha=1}^{p} \sum_{i=1}^{c} \mu_i^\alpha\, dN_i^\alpha = 0 \tag{V.8}$$

for reversible changes of S and V. However, in the closed system $\sum_{\alpha=1}^{p} dN_i^\alpha = 0$, or

$$\sum_{\alpha=2}^{p} \sum_{i=1}^{c} (\mu_i^\alpha - \mu_i^1)\, dN_i^\alpha = 0\,. \tag{V.9}$$

In this equation the dN_i^α's are independent and thus, if the sum is to vanish, $\mu_1^\alpha = \mu_i^1$ for all α, i.e., the chemical potentials of each component is the same in all phases.

V.6 Components and Species, Chemical Equilibrium

Components, however, are not necessarily the chemical substances of interest in a system. If, for example, a system is made from a "recipe" which says take H_2O and raise its temperature to 2500 K and fix its pressure at 10^{-6} bar, then the amount of molecular H_2O that remains will be negligible, and the identifiable species in the system will include predominantly other members in the set: H, O, OH, H_2, O_2, H_2O. Among these only a certain number will have independently variable mole numbers.

In order to obtain the restraints that equilibration places upon the variation in quantities of chemical species in a system consider first equilibration within a single phase, e.g., the vapor phase. We first consider the states that would be accessible to the system if the moles of each species could be varied independently, i.e., perform a thought experiment in which each species is a component with respect to which the system is open, and the chemical reactions are prevented from occurring. All considered changes are reversible. For this case:

$$dU = T\, dS - P\, dV + \sum_{i=1}^{s} \mu_i\, dN_i\,. \tag{V.10}$$

Next consider the closed system in which all reactions occur reversibly. For this case:

$$dU = T\, dS - P\, dV\,, \tag{V.11}$$

because the system is closed and thus $dN_i = 0$ for all components. Now if the system in the open system changes through exactly the same states as in the closed system case, i.e., if we set

$$dN_i = \sum_{j=1}^{r} \nu_i^j\, d\xi_j \tag{V.12}$$

where ξ_j is the extent of progress of the jth reaction in which the stoichimetric coefficient of the ith species is ν_i^j, then, combining the above,

$$\sum_{i=1} \sum_{j=1} \mu_i \nu_i^j\, d\xi_j = 0\,, \tag{V.13}$$

and since the reactions are independent, the coefficient of $d\xi_j$ must vanish for each j, i.e.,

$$\sum_{i=1} \mu_i \nu_i^j = 0 \qquad \text{(V.14)}$$

for each reaction.

Thus there is, for each of the r independent chemical reaction with respect to which the s chemical species are equilibrated, one thermodynamic restriction (which ultimately translates into an equilibrium constant). Hence the number of independent chemical content variables in a system which is otherwise (i.e., other than by equilibration of chemical reactions among species) unrestricted is $c = s - r$. It should be noted that the prescribed recipe for formation of the system might include "arbitrary" (determined by the recipe) restraints.

For example in the case of the recipe: take 10 grams of H_2O to $T = 2500$ K and $P = 10^{-6}$ bar, where the species are determined to be: H, O, OH, O_2, H_2, H_2O, and a set of independent uniquely balancable (characterized by a single extent) chemical reactions is:

$$H + O = OH \qquad \text{(V.15)}$$

$$2\,O = O_2 \qquad \text{(V.16)}$$

$$2\,H = H_2 \qquad \text{(V.17)}$$

$$2\,H + O = H_2O \qquad \text{(V.18)}$$

it follows that in the unrestricted system $c_{unres} = 6 - 4 = 2$. However, the recipe has arbitrarily fixed the overall H to O ratio at 2:1 (because the recipe specified taking H_2O) and thus

$$N_H + N_{OH} + 2\,N_{H_2} = 2(N_O + N_{OH} + 2\,N_{O_2})\,. \qquad \text{(V.19)}$$

Furthermore since this equation implies a relationship among **intensive** variables (e.g., divide the above by $N_{total} = N_H + N_O + N_{OH} + N_{O_2} + N_{H_2O}$ to obtain

$$X_H + 2X_{H_2} = 2X_O + X_{OH} + 4X_{O_2}\,, \qquad \text{(V.20)}$$

a relationship among the mole fractions) it must be the case that the number of intensive variables is less by one than would be the case if the system were unrestricted.

This leads to a slightly modified form of the phase rule

$$f = (c_{unres} - \varrho) - p + 2 \qquad \text{(V.21)}$$

where, if you wish you may, somewhat arbitrarily, take $c = c_{unres} - \varrho$ and write

$$f = c - p + 2\,. \qquad \text{(V.22)}$$

While identifying $c_{unres} - p$ with the number of components makes complete sense in the case just considered, since the recipe specified one chemical substance and the conclusion reached is that $c = 2 - 1 = 1$, this identification is not always so clear cut.

For example, if a metal sulfate MSO_4 were decomposed to yield a molten metal solution with dissolved S and O (MS_vO_v) and gaseous SO and SO_2, it would follow that (letting N^l_M = moles of species M in the liquid, N^l_S = moles of species S in the liquid, N^v_{SO} = moles of species SO in the vapor, etc.):

$$N^l_M = N^l_S + N^v_{SO} + N^v_{SO_2} \tag{V.23}$$

and

$$4N^l_M = N^l_O + N^v_{SO} + 2N^v_{SO_2} \tag{V.24}$$

or, collecting terms in the liquid and vapor on each side of each equation and dividing

$$\frac{N^l_M - N^l_S}{4N^l_M - N^l_O} = \frac{N^v_{SO} + N^v_{SO_2}}{N^v_{SO} + 2N^v_{SO_2}}, \tag{V.25}$$

and, dividing the numerator and denominator, on the left side by $N^l_M + N^l_S + N^l_O$ and on the right side by $N^v_{SO} + N^v_{SO_2}$:

$$\frac{4X^l_M - X^l_S}{4X^l_M - X^l_O} = \frac{X^v_{SO} + X^v_{SO_2}}{X^v_{SO} + 2X^v_{SO_2}}, \tag{V.26}$$

i.e., there is in the recipe an arbitrary restriction upon intensive variables. The species: M, S, O, SO and SO_2 allow the following independent reactions

$$S(soln) + SO_2(g) = 2\ SO(g) \tag{V.27}$$

and

$$O(soln) + SO(g) = SO_2(g)\,, \tag{V.28}$$

thus $c_{unres} = 5 - 2 = 3$. Since $\varrho = 1$, $c_{unres} - \varrho = 2$, and $f = 2 - 2 + 2 = 2$. There does not appear in this case to be any straightforward rationale for saying that this is a two component system, although it is, of course, correct to say that it is a system with two independent composition variables. If it is desired to maintain the component terminology it is perhaps preferable to refer to the system as a three component system with an arbitrary restriction upon the intensive variables.

The distinction between moles of component (moles of chemical substance added to a system from the surroundings, N^C_i) and moles of species (moles of chemical substance actually present according to measurement, N^s_i) raises a question concerning chemical potential, μ_i, namely is $\mu^s_i = \left(\frac{\partial U}{\partial N^s_i}\right)_{T,P,N^s_j}$ or $\mu^c_i = \left(\frac{\partial U}{\partial N^c_i}\right)_{T,P,N^c_j}$ the appropriate quantity? The answer is that they are equal for a system at chemical equilibrium, i.e., the slope of the internal energy vs moles of ith chemical substance added at constant entropy, volume and N_j is the same whether N_j is exchanged with the surrounding (regardless of its ultimate chemical fate, e.g., essentially total decomposition in the case of H_2O in the system at 2500 K and 10^{-6} bar) or is taken to be the moles of i actually present in the system as determined by the usual methods of species identification (optical spectroscopy, NMR, ESR, mass spectrometry, etc).

In order to see that this is so consider the case of

$$dU = T\,dS - P\,dV + \sum_{i=1}^{s} \mu_i^s\,dN_i^s\,, \tag{V.29}$$

the Gibbsian equation applied to a homogeneous mixture of species in which a reaction (thought experiment) is restrained from occurring and the system is open to the exchange of all species via reversible processes, and compare this with

$$dU = T\,dS - P\,dV + \sum_{i=1}^{c} \mu_i^c\,dN_i^c \tag{V.30}$$

considering only the states accessible to the system that is open with respect to components.

An example is helpful here. Let the components be $H_2(g)$ and $O_2(g)$, and the species be $H_2(g)$, $O_2(g)$ and $H_2O(g)$ with the chemical reaction being

$$H_2(g) + \tfrac{1}{2}\,O_2(g) = H_2O(g)\,. \tag{V.31}$$

Now by oxygen and hydrogen balance, since the states are those accessible with components H_2 and O_2,

$$N_{H_2}^c = N_{H_2}^s + N_{H_2O}^s \tag{V.32}$$

and

$$N_{O_2}^c = N_{O_2}^s + \tfrac{1}{2}\,N_{H_2O}^s \tag{V.33}$$

and thus

$$dN_{H_2}^c = dN_{H_2}^s + dN_{H_2O}^s \tag{V.34}$$

and

$$dN_{O_2}^c = dN_{O_2}^s + \tfrac{1}{2}\,dN_{H2O}^s \tag{V.35}$$

or

$$dN_{H_2}^s = dN_{H_2}^c - dN_{H_2O}^s \tag{V.36}$$

and

$$dN^s_{O_2} = dN^c_{O_2} - \tfrac{1}{2} dN^s_{H_2O} \,. \tag{V.37}$$

Substituting into Eq. (V.29)

$$dU = T\,dS - P\,dV + \mu^s_{O_2}\left(dN^c_{O_2} - \frac{1}{2}\,dN^s_{H_2O}\right)$$

$$+ \mu^s_{H_2}(dN^c_{H_2} - dN^s_{H_2O}) + \mu_{H_2O}\,dN^s_{H_2O} \tag{V.38}$$

or

$$dU = T\,dS - P\,dV + \mu^s_{O_2}\,dN^c_{O_2} + \mu^s_{H_2}\,dN^c_{H_2}$$

$$+ \left(\mu^s_{H_2O} - \mu^s_{H_2} - \frac{1}{2}\,\mu^s_{O_2}\right) dN^s_{H_2O} \tag{V.39}$$

where the coefficient of $dN^s_{H_2O}$ is $\Sigma\, \nu_i \mu_i$ for reaction IV.31 and is zero for reversible changes of state. Thus

$$dU = T\,dS - P\,dV + \mu^s_{O_2}\,dN^c_{O_2} + \mu^s_{H_2}\,dN^c_{H_2} \tag{V.40}$$

and since

$$dU = T\,dS - P\,dV + \mu^c_{O_2}\,dN^c_{O_2} + \mu^c_{H_2}\,dN^c_{H_2}\,, \tag{V.41}$$

the independence of the variables S, V, $N^c_{O_2}$ and $N^c_{H_2}$ yields $\mu^s_{O_2} = \mu^c_{O_2}$ and $\mu^s_{H_2} = \mu^c_{H_2}$.

V.7 Some Equations of Use in the Consideration of Heterogeneous Equilibria

The Gibbsian equation for U can be integrated at constant intensive state:

$$\int dU = T \int dS - P \int dV + \sum_{i=1}^{c} \mu_i \int dN_i \tag{V.42}$$

to yield

$$U = TS - PV + \sum_{i=1}^{c} \mu_i N_i \,. \tag{V.43}$$

Differentiation of this equation yields

$$dU = T\,dS + S\,dT - P\,dV - V\,dP + \sum_{i=1}^{c} \mu_i\,dN_i + \sum_{i=1}^{c} N_i\,d\mu_i \tag{V.44}$$

which by comparison with the Gibbsian equation for U yields the Gibbs-Duhem equation

$$S\,dT - V\,dP + \sum_{i=1}^{c} N_i\,d\mu_i = 0\,. \tag{V.45}$$

Application of this equation to a c = 1, two phase (α and β) system with $T^\alpha = T^\beta$, $P^\alpha = P^\beta$ and $\mu^\alpha = \mu^\beta$ (i.e., at thermal, mechanical and chemical equilibrium without constraints) yields

$$\bar{S}^\alpha\,dT - \bar{V}^\alpha\,dP + d\mu^\alpha = 0 \tag{V.46}$$

and

$$\bar{S}^\beta\,dT - \bar{V}^\beta\,dP + d\mu^\beta = 0\,, \tag{V.47}$$

where $\bar{S}^\alpha = S^\alpha/N^\alpha$, $\bar{V}^\alpha = V^\alpha/N^\alpha$, etc., and $d\mu^\alpha = d\mu^\beta$ for the p = 2 equilibrium states. Substraction yields

$$\Delta\bar{S}\,dT - \Delta\bar{V}\,dP = 0\,, \tag{V.48}$$

or

$$\frac{dP}{dT} = \frac{\Delta\bar{S}}{\Delta\bar{V}} \tag{V.49}$$

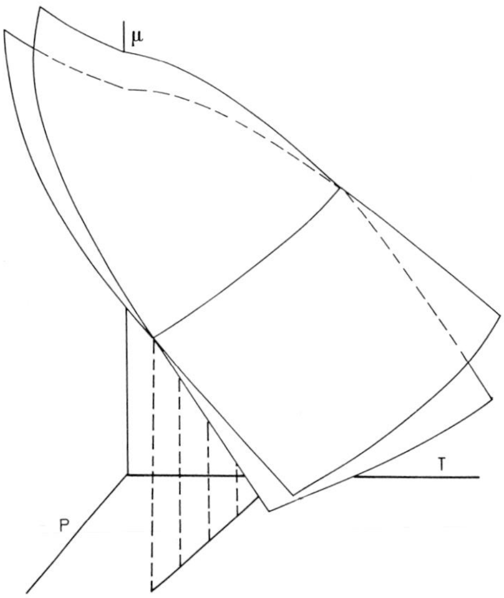

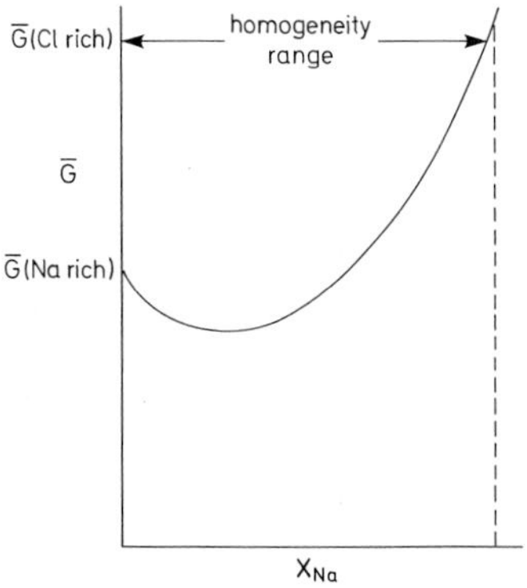

Fig.V.2. Schematic $\overline{G}$ vs X across a homogeneity range.

Fig. V.1. The single component T—P phase coexistence line as the projection of the intersection line of two μ (or $\overline{G}$) surfaces.

This is the differential equation of the coexistence line for c = 1, p = 2 system in T-P space providing $\Delta V \neq 0$. The $\Delta V = 0$ case, second-order phase transition, is the subject of Chap. VII.

The T-P curve is the line in T-P space along which $\mu^\alpha = \mu^\beta$. It is instructive to consider $\mu^\alpha(T, P)$ and $\mu^\beta(T, P)$ surfaces (Fig. V.1) and to note that the T-P coexistence lines (phase diagram lines) are projections of the line of intersection of μ^α and μ^β onto the T-P plane.

The Gibbsian equation with which we have dealt up to this point is one which follows naturally from an application of the first and second laws of thermodynamics to reversible changes of state in open systems. The independent variables (S, V, N_i) are all extensive. It is possible to define new functions such that some of the natural (natural here means arising from a transformation of the Gibbsian equation for dU) independent variables are intensive. That is since $U = TS - PV + \sum_{i=1}^{c} \mu_i N_i$ we can define H (the enthalpy), A (the Helmholz Free Energy) and G (the Gibbs Free Energy)

$$H = U + PV = TS + \sum_{i=1}^{c} \mu_i N_i \tag{V.50}$$

$$A = U - TS = -PV + \sum_{i=1}^{c} \mu_i N_i \tag{V.51}$$

$$G = U - TS + PV = \sum_{i=1}^{c} \mu_i N_i \tag{V.52}$$

and obtain

$$dH = dU + P\,dV + V\,dP = V\,dP + T\,dS + \sum_{i=1}^{c} \mu_i\,dN_i \tag{V.53}$$

$$dA = dU - T\,dS - S\,dT = P\,dV - S\,dT + \sum_{i=1}^{c} \mu_i\,dN_i \tag{V.54}$$

$$dG = dU - T\,dS - S\,dT + P\,dV + V\,dP = V\,dP - S\,dT + \sum_{i=1}^{c} \mu_i\,dN_i \tag{V.55}$$

and we see that, in terms of their "natural" independent variables, $H = H(P, S, N_i)$, $A = A(V, T, N)$ and $G = G(T, P, N)$. It also follows that

$$\left(\frac{\partial H}{\partial N_i}\right)_{P,S,N_j} = \left(\frac{\partial A}{\partial N_i}\right)_{T,V,N_j} = \left(\frac{\partial G}{\partial N_i}\right)_{T,P,N_j} = \left(\frac{\partial U}{\partial N_i}\right)_{S,V,N_j} \tag{V.56}$$

This last equation, namely that equating the chemical potential to the partial derivative of G, the Gibbs free energy, with respect to N_i at constant temperature, pressure and N_j is one that is frequently used. It is because of the special role that constant temperature and pressure play in our experiments and thinking that this equation is found to be so useful.

V.8 Thermodynamics of Nonstoichiometry

Consider a two component solid with variable composition. For such a solid (by V.52)

$$G = \sum_{i=1}^{c} N_i \mu_i = N_1 \mu_1 + N_2 \mu_2 . \tag{V.57}$$

In order to obtain an exaggerated view of what is being discussed in this section think of NaCl, for which the variation in composition is on the order of parts per million. When NaCl coexists in equilibrium with Na(s) it is saturated with Na. In order to think about μ_{Na} recall that μ_{Na} (in NaCl) $= \mu_{Na}$ (in Na) where the latter is very nearly (since the Na(s) in equilibrium with NaCl is almost pure Na(s)) the value appropriate for equilibrium with Na(g) at the vapor pressure of Na(s) (P^0)

$$\mu_{Na} = \mu_{Na}^{0,g} + RT \ln P_{Na}^0 . \tag{V.58}$$

This value of $P_{Na} (= P_{Na}^0)$ is many orders of magnitude greater than the value of P_{Na} over pure, stoichiometric NaCl, and is also many (actually a very great many) orders of magnitude greater than P_{Cl_2}. Note that we can write the reaction

$$NaCl(s) = Na(g) + \tfrac{1}{2} Cl_2(g) \tag{V.59}$$

and thus, since the NaCl is virtually unchanged (only by ppm) in going from stoichiometric NaCl(s) to NaCl(s) saturated with Na, we have

$$K \cong P_{Cl_2}^{1/2} P_{Na} \tag{V.60}$$

and thus a greatly increased P_{Na} implies a greatly diminished P_{Cl_2}.

On the other hand, the situation is reversed when NaCl is in equilibrium with Cl_2 at a pressure of 1 atm. Thus we can see that the chemical potential of Na (the slope of G vs N_{Na} at constant T, P, N_{Cl_2}) varies by a very large amount across the NaCl(s) homogeneity range (small though it is). One minor but sometimes disturbing point should be dismissed, namely the effect of the change in applied (hydrostatic) pressure upon the chemical potentials. This effect can be estimated from

$$\left(\frac{\partial \mu_i}{\partial P}\right)_T = \bar{V}_i ,$$

where $\bar{V}$ is the partial molar volume, and this estimate shows that for pressures less than 1 bar the effect is entirely negligible. That is, so far as our discussion is concerned, the entire range of homogeneity of NaCl discussed is at zero applied pressure so far as the dependence of μ_i in the solid is concerned.

For this reason it is possible to consider a $\overline{G} = G/(N_{Na} + N_C)$ vs $X_{Na} = N_{Na}/(N_{Na} + N_C)$ curve at "constant" pressure and temperature across the NaCl solid solution range (from equilibrium with Na(s) to, arbitrarily, equilibrium with $Cl_2(g)$ at $P_{Cl_2} = 1$ bar). Consider any two compositions $NaCl_a$ and $NaCl_b$ (with $a \neq b$ although both are necessarily close to 1). For the reaction

$$\frac{1}{1+m} NaCl_a + \frac{m}{1+m} NaCl_b = NaCl_{\frac{1a+mb}{1+m}} \qquad (V.61)$$

$\Delta G < 0$ because, by the definition of a solid solution range, such a process is spontaneous at constant temperature and pressure. Thus (letting $c = \frac{1+mb}{1+m}$):

$$\overline{G}(NaCl_c) < \frac{1}{1+m} \overline{G}(NaCl_a) + \frac{m}{1+m} \overline{G}(NaCl_b) . \qquad (V.62)$$

However, $\frac{1}{1+m} \overline{G}(NaCl_a) + \frac{m}{1+m} \overline{G}(NaCl_b)$ defines a straight line connecting $\overline{G}(NaCl_b)$ with $\overline{G}(NaCl_a)$ on a plot of $\overline{G}$ vs X_{Na}, and thus $\overline{G}(NaCl_c)$ falls below this line for all 1, m. In other words $\overline{G}$ vs X_{Na} is concave up across the $NaCl_x$ homogeneity range (Fig. V.2).

Now, shifting the discussion to some arbitrary metal (M) — nonmetal (Y) system and recognizing that there may be more than one compound present in the system yields, for example, $\overline{G}$ vs X_Y as in Fig. V.3.

For any point on any curve:

$$G = N_M\mu_M + N_Y\mu_Y , \qquad (V.63)$$

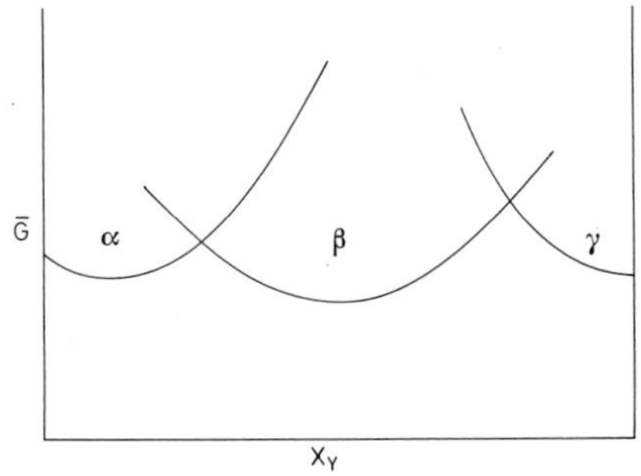

Fig. V.3

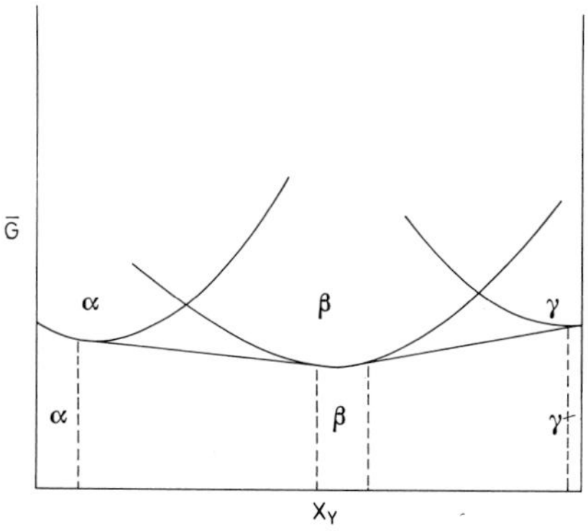

Fig. V.4

Fig. V.3. Schematic $\overline{G}$ vs X_Y for β stable wrt disproportionation into α and γ.

Fig. V.4. Illustrating that heterogeneous equilibrium implies equal slopes of $\overline{G}$ vs X and relationship to homogeneity ranges of condensed phases.

or

$$\overline{G} = X_M\mu_M + X_Y\mu_Y\,. \tag{V.64}$$

Taking the derivative with respect to X_M and making use of $\frac{dX_M}{dX_Y} = -1$:

$$\left(\frac{\partial\bar{G}}{\partial X_M}\right)_{T,P} = \mu_M - \mu_Y + X_M\left(\frac{\partial\mu_M}{\partial X_M}\right)_{T,P} + X_Y\left(\frac{\partial\mu_Y}{\partial X_M}\right)_{T,P}. \tag{V.65}$$

By the Gibbs-Duhem equation

$$X_M\left(\frac{\partial\mu_M}{\partial X_M}\right)_{T,P} + X_Y\left(\frac{\partial\mu_Y}{\partial X_M}\right)_{T,P} = 0\,, \tag{V.66}$$

and thus

$$\left(\frac{\partial\bar{G}}{\partial X_M}\right)_{T,P} = \mu_M - \mu_Y\,. \tag{V.67}$$

Since when two phases coexist in equilibrium $\mu_M^\alpha = \mu_M^\beta$ and $\mu_Y^\alpha = \mu_Y^\beta$, it follows from Eq. V.67 that the slopes of their $\overline{G}$ vs X_M curves are equal. Thus we can find the limits of homogeneity by constructing the common tangents as shown on Fig. V.4.

It is important to note that all compounds exhibit solid solubility (even NaCl) since the chemical potential of, for example, M must differ when α coexists with M(s) and when α coexists with β. This solid solubility may be for many purposes negligible (as with NaCl) or may be very wide (as with TiO where the range is on the order of 0.3). It is also important to note that the range of homogeneity (e.g., for β in Fig. V.4) depends not only upon the chemistry and physics of β but also upon those of α and γ. One feature that determines the width of the β homogeneity range is the steepness of G vs X curve, and others are the relative locations of the α and γ curves and their slopes.

V.9 Stability and Instability with Respect to Disproportionation

It could occur that the β curve lies above the common tangent to α and γ as shown in Fig. V.5. In this case β is unstable with respect to disproportionation and β cannot be formed by reaction of α and γ. It can perhaps be formed as a metastable compound via some alternative route, but α and γ will not react to form even a small amount of β. Since we are accustomed to thinking about reactions in vapor or solution phases we are also accustomed to thinking about an equilibrium constant which implies the existence of a small amount of product even for strongly unfavored reactions. However, the situation in vapor or the solution cases is that the chemical potentials of reactants and products change until $\sum_i \nu_i\mu_i = 0$. For solid phase reactions such as $\alpha + \gamma = \beta$ in the

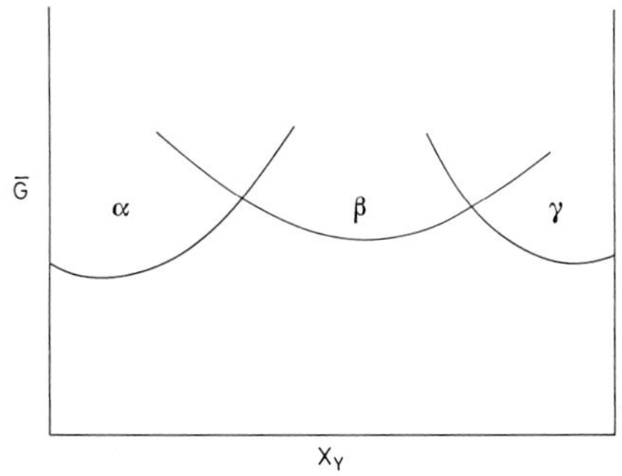

Fig. V.5

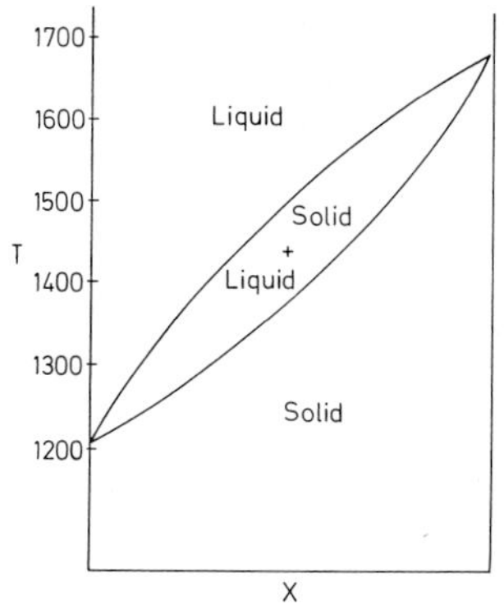

Fig. V.6

Fig. V.5. Schematic $\overline{G}$ vs X_Y illustrating β unstable wrt disproportionation.
Fig. V.6. Phase diagram (T—X) for a perfect liquid solution-perfect solid solution equilibrium.

case of Fig. V.5, no variation of composition will alter the fact that for $\beta = \alpha + \gamma$, $\Delta G < 0$, and the process is therefore spontaneous.

It sometimes happens that the situation of Fig. V.5 changes to that of Fig. V.4 with, for example, changes in temperature at constant pressure. There then exists at a given pressure some T at which the α, β and γ curves have a common tangent. At this temperature, T′, α, β and γ all coexist in equilibrium. Suppose β is a high-temperature phase, i.e., that for $T' + \delta T$ the situation is that of Fig. V.4 (β stable) and for $T' - \delta T$ the situation is that of Fig. V.5 (β unstable). Then for $\beta = \alpha + \gamma$

$$\Delta G(T' + \delta T) = \Delta H_{T'} - (T' + \delta T)\,\Delta S_{T'} > 0 \qquad \text{(V.68)}$$

and since α, β and γ coexist at equilibrium when $T = T'$,

$$\Delta G(T') = \Delta H_{T'} - T'\,\Delta S_{T'} = 0\,. \qquad \text{(V.69)}$$

Combining the above

$$-\delta T\,\Delta S_{T'} > 0 \qquad \text{(V.70)}$$

or $\Delta S_{T'} < 0$. I.e., in order that a compound be stable with respect to disproportionation at $T' + \delta T$ and unstable at $T' - \delta T$ it is necessary that the compound disproportionate with a decrease in entropy (form from adjacent phases with an increase in entropy). Furthermore, since $\Delta H_{T'} = T'\,\Delta S_{T'}$, it follows that such a high temperature phase forms endothermically. If it had been assumed that the phase disproportionated with increasing temperature the opposite result would have been obtained, i.e., the compound would be found to form from adjacent phases exothermically and exoentropically. Thus phases which form from their neighbors by $p = 3$ solid state reactions with decreasing temperature are "normal" in the sense that they form because they are energetically stable.

On the other hand, phases which form from adjacent phases with increasing temperature are "abnormal" in the sense that they are high energy phases which are entropically stable.

V.10 Thermodynamics of Defects

As has been emphasized, one important aspect of the application of thermodynamics to solids is the treatment of compounds as solid solutions. This treatment is appropriate for "line" compounds with extremely narrow ranges of homogeneity because of the extremely large variations of chemical potential that can accompany the otherwise negligible (parts per million) changes in chemical composition. The treatment is, on the other hand, essential rather than appropriate to the understanding of the behavior of some compounds which exhibit very wide ranges of homogeneity. In such cases the very meaning of the terms solution and compound become unclear and it is therefore not possible to draw a sharp distinction between these terms and a meaningful discussion can only be carried out in terms of phases.

However, it may well be the case that a particular set of properties, e.g., chemical inertness in a given environment or electrical conductivity properties of a given character, depend even for "line compounds" upon the stoichiometry in an important way. For example, it frequently is found that hole and electron conductivities increase markedly with, respectively, increases and decreases in nonmetal-to-metal ratios in solids (because of the increases in the concentrations of holes and conduction electrons). For example, consider the reaction of a metal-nonmetal (MX) compound with gaseous diatomic $X_2(g)$:

$$\tfrac{1}{2}\,X_2(g) = V_M + X^{2-} + 2\,h\,(V.B.)\,. \qquad (V.71)$$

The meaning of this reaction is that gaseous X_2 is in equilibrium with the compound MX, and if this equilibrium is shifred to the right (e.g., by increasing the X_2 pressure), the shift occurs with the formation of a vacancy on the metal lattice, a nonmetal on the nonmetal lattice with, on the average a very nearly filled p shell, and the removal of two electrons from the delocalized valence band states.

For cases for which reaction V.71 is appropriate, and for which the stoichiometric MX compound has $[V_M] = 0$ ($[V_M]$ = concentration of metalatom sites vacant),

$$[h] = 2[V_M] \qquad (V.72)$$

and

$$K = \frac{[V_M]\,[h]^2}{P_{X_2}^{1/2}}\,, \qquad (V.73)$$

provided $[X^{2-}]$ does not vary significantly, or

$$2K = \frac{[h]^3}{P_{X_2^{1/2}}} . \tag{V.74}$$

Hence the concentration of holes, upon which the hole conductivity in part depends, is proportional to the $\frac{1}{6}$ power of the X_2 pressure and therefore as the X/M ratio is increased by increasing P_{X_2}, the hole conductivity increases as well. On the metallic conductivity side, again taking P_{X_2} as the controlled variable,

$$X^{2-} = \tfrac{1}{2} X_2(g) + V + 2e^-(\text{C.B.}) \tag{V.75}$$

i.e., the shift in equilibrium with diminished P_{X_2} produces additional electrons in the conduction band. In cases to which reactions V.71 and V.75 apply it is sometimes the aim to produce "intrinsic" MX, i.e., stoichiometric MX with $[e^-] = [h] \cong 0$ and only ionic conductivity. For the synthesis of an intrinsic material it is essential to understand the nonstoichiometric nature of solids and the heterogeneous equilibria in which they can be involved.

V.11 Phase Diagrams

Phase diagrams provide insight into such heterogeneous equilibria. One of the simplest phase diagrams that can occur is that for a two phase equilibrium between two perfect solutions, i.e., two solutions in which there are no energy effects and only the effects of randomization of the dissolving species. In the case of solids this means random on the average over a region of the solid, since at any site at any time the occupation is, of course, not by an average species, but is, of course, by M, M′, V, etc.

At any rate the perfect solution chemical potential is given by

$$\mu_1 = \mu_1^0 + RT \ln X_1 \tag{V.76}$$

where X_1 is the mole fraction of component 1 in the solution, and μ_1^0 is the chemical potential of pure component 1. Since we are considering equilibria between two perfect solutions (α and β), and equilibrium implies that $\mu_1^\alpha = \mu_1^\beta$, we have

$$\mu_1^{0,\alpha} + RT \ln X_1^\alpha = \mu_1^{0,\beta} + RT \ln X_1^\beta \tag{V.77}$$

or

$$-RT \ln \frac{X_1^\alpha}{X_1^\beta} = \Delta\mu^0 . \tag{V.78}$$

The mole fractions on the left-hand side of this expression refer to the two solutions coexisting at the temperature T, the right hand side, on the other hand, is determined by the difference of chemical potentials between two pure phases α and β. Dividing by T, taking the derivative with respect to T^{-1} and recognizing that

$$\frac{d\,\Delta\mu^0/T}{dT^{-1}} = \Delta H_1^0\,, \tag{V.79}$$

where ΔH_1^0 is the enthalpy of fusion of pure 1,

$$\frac{d\ln\dfrac{X_1^\alpha}{X_1^\beta}}{dT^{-1}} = \frac{\Delta H_1^0}{R} \tag{V.80}$$

which can be integrated. The assumption that ΔH_1^0 is independent of T yields the approximation

$$\ln\frac{X_1^\alpha}{X_1^\beta} = -\frac{\Delta H_1^0}{R}\left\{\frac{1}{T} - \frac{1}{T_1}\right\} \tag{V.81}$$

where T_1 is the temperature at which $X_1^\alpha = X_1^\beta = 1$, i.e., at which pure α and β coexist in equilibrium. Similarly for component 2

$$\ln\frac{X_2^\alpha}{X_2^\beta} = -\frac{\Delta H_2^0}{R}\left\{\frac{1}{T} - \frac{1}{T_2}\right\}, \tag{V.82}$$

and using $X_1^\alpha + X_2^\alpha = 1$ and $X_1^\beta + X_2^\beta = 1$ provides the T—X diagram shown in Fig. V.6.

The phase diagram of Fig. V.6 seldom is even approximately correct for condensed phase — condensed phase equilibria. It results from the essentially impossible case in which the energetics of the interaction of component 1 with component 2 in both the α and the β phase are the same as those of component 1 with itself and of component 2 with itself as well as random distributions (no short range ordering) in the condensed phases.

For the usual cases in which there are nonnegligible differences in the energies of interaction, other terms can arise in the chemical potential. The simplest of these is regular solution model which takes the form

$$\mu_1 = \mu_1^0 + RT\ln X_1 + WX_2^2\,. \tag{V.83}$$

By the Gibbs-Duhem equation it then results that

$$\mu_2 = \mu^0{}_2 + RT\ln X_2 + WX_1^2\,, \tag{V.84}$$

In this case

$$G_{soln} = N_1\mu_1 + N_2\mu_2 \tag{V.85}$$

$$= N_1\mu_1^0 + N_2\mu_2^0 + N_1RT\ln X_1 + N_2RT\ln X_2 + W(N_1X_2^2 + N_2X_1^2)\,. \tag{V.86}$$

Dividing by $N_1 + N_2$ and identifying $\overline{G}_{soln} - X_1\mu_1^0 - X_2\mu_2^0$ as $\Delta\overline{G}_{soln}$,

$$\Delta\overline{G}_{soln} = RT(X_1 \ln X_1 + X_2 \ln X_2) + W(X_1X_2^2 + X_2X_1^2)\,. \tag{V.87}$$

Since $\left(\frac{\partial \Delta\overline{G}_{soln}}{\partial T}\right)_P = -\Delta\overline{S}_{soln}$ we find

$$\Delta\overline{S}_{soln} = -R(X_1 \ln X_1 + X_2 \ln X_2)\,, \tag{V.88}$$

the perfect solution value, and then since $\Delta\overline{G} = \Delta\overline{H} - T\,\Delta\overline{S}$ we have

$$\Delta\overline{H}_{soln} = W(X_1X_2^2 + X_2X_1^2) \tag{V.89}$$

$$= WX_1X_2\,. \tag{V.90}$$

Hence the regular solution has a nonzero enthalpy of mixing as shown in Fig. V.7.

One feature of regular solution behavior of relevance to the behavior of real systems is that the heterogeneous equilibrium between two regular solution phases can exhibit the more usually observed maximum (Fig. V.8) or minimum, congruent phase transition at some intermediate temperature. At such a transition point $\mu_1^\alpha = \mu_1^\beta$ and

$$\mu_1^{0,\alpha} + RT \ln X_1^\alpha + W^\alpha(X_2^\alpha)^2 = \mu_1^{0,\beta} + RT \ln X_1^\beta + W^\beta(X_2^\beta)^2\,, \tag{V.91}$$

$X_1^\alpha = X_1^\beta$ and $X_2^\alpha = X_2^\beta = X_2$ and thus

$$\Delta\mu_1^0 = X_2^2\,\Delta W\,. \tag{V.92}$$

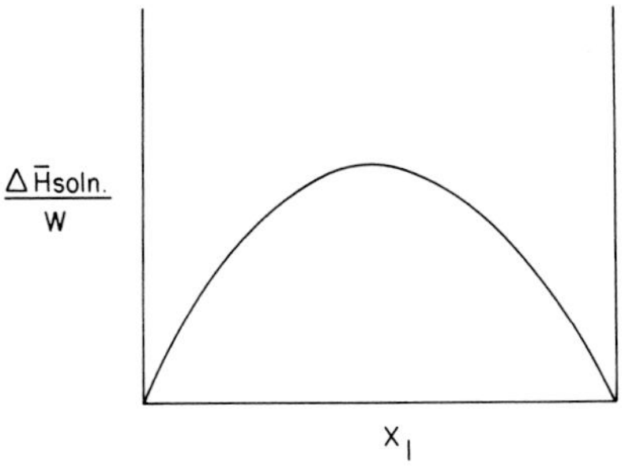

Fig. V.7. Molar enthalpy of solution for a regular solution.

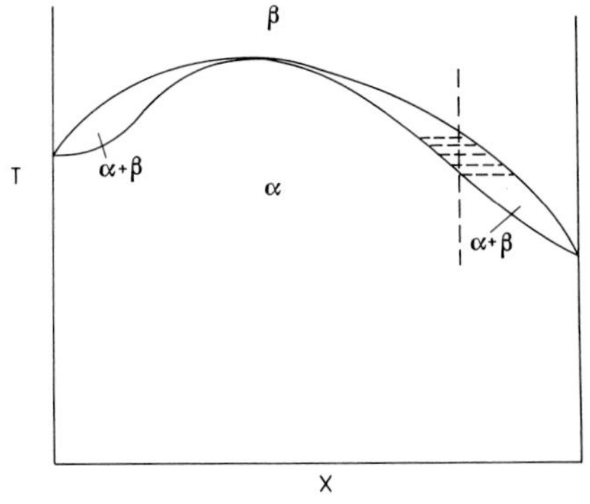

Fig. V.8. Schematic binary 1-X diagram with a congruent transition.

Furthermore, $\Delta\mu_1^0 = \Delta H_1^0 - T_c\,\Delta S_1^0$ (T_c = congruent equilibrium temperature) if $\Delta C^0_{p,1} = 0$, and thus

$$\Delta H_1^0 - T_c\,\Delta S_1^0 = X_2^2\,\Delta W\,. \tag{V.93}$$

Similarly

$$\Delta H_2^0 - T_c\,\Delta S_2^0 = X_1^2\,\Delta W\,. \tag{V.94}$$

These two equations serve to fix T_c and X for a given ΔW. Of course real systems will not in general conform to the simple regular solution behavior, however, the point here is that even such a simple assumption of deviation from perfect solution behavior can lead to the complexity of a maximum or minimum congruent transformation.

The relevance of such behavior to solid-state synthesis can be seen as follows. When β (Fig. V.8) which might be solid or liquid, is cooled from high temperature it may be well annealed and therefore well characterized. However, upon proceeding through the α-β coexistence region, unless it does so through the congruent point, the system will separate into two phases with different compositions at different temperatures. If it is the case that the temperature change is too rapid for α to remain uniform at the composition appropriate to each, then the solid will not be homogeneous. That is, if the coexistence temperatures are sufficiently low that diffusion times are long relative to times of significant temperature change then α may form from well characterized β in a poorly characterized, compositionally non-uniform aggregate.

V.12 Defect Ordering

Within a solid solution region there are sites which might or might not be occupied or might be occupied by element A or element B. In either case it follows that there is a configurational contribution to the entropy of the solid solution, i.e., that for solid solutions at high temperatures a configurational entropy term will contribute to the stability of the defect solid. In general there will be a competing energetic term which tends to stabilize some ordering of the solid and this competition will lead to a tendency for wide ranges of homogeneity at high temperatures to break up into smaller, more ordered (lower configurational entropy) ranges with decreasing T. For this reason it is anticipated that wide ranges of homogeneity at high temperatures will, through order-disorder processes, heterogeneous reactions, etc. subdivide at lower temperatures into narrower ranges or more nearly stoichiometric compounds. An interesting example of this behavior is provided by the Cr—S system which is dicussed in detail in Sect. IX.7.

V.13 Problems

1. How many independent intensive variables are there in the system containing NH_3(g), NH_2(g), NH(g), N_2(g), N(g), H_2(g) and H(g) formed

a. by heating a mixture of $N_2(g)$ and $H_2(g)$ to high temperature,
b. by heating $NH_3(g)$ to high temperature?

2. A metal oxychloride, $MOCl_2$, is heated to very high temperatures at which the vapor contains $O_2(g)$, O(g), OCl(g), $Cl_2(g)$ and Cl(g) and the condensed phase consists of M with small amounts of O and Cl in solid solution. How many independent intensive variables are there? How many components? Give your reasoning.
3. Derive the condition for chemical equilibrium for a single homogeneous reaction using $dA = -S\,dT - P\,dV + \Sigma\,\mu_i\,dN$.
4. Given the data tabulated below calculate the partial pressures of $Cl_2(g)$, Cl(g) and Na(g) over NaCl in equilibrium with Na (taken to be pure) and in equilibrium with chlorine gas at P = 1 bar. at 298 K.

	$\Delta H_f^0(298)$ kJ/mol	$S^0(298)$ J/deg. mol
Na(s)		51.2
Na(g)	107.32	153.603
NaCl(s)	−411.153	72.13
$Cl_2(g)$		222.957
Cl(g)	121.679	165.088

5. Given the following melting temperatures and enthalpies of fusion for Si ($T_m = 1685$ K, $\Delta H_f^0 = 46$ kJ (mole) and for Ge ($T_m = 1213$ K, $\Delta H_f^0 = 34.7$ kJ/mole) plot the T—X phase diagram for Ge—Si assuming perfect solid solution and perfect liquid solution behavior and compare with the literature result (Stöhr, H., Klemm, W.: Z. Anorg. Chem. **241**, 305 (1939)).
6. If the chemical potential of component A of a solid solution is given by $\mu_A = \mu_A^0 + RT \ln X_A + \alpha X_B^2 + \beta X_B^3$ find the form for the chemical potential of B in the solution taking pure B as the standard state. Comment on the validity of the result if A and B are not recognizable species in the solid solution.
7. Using the data given below determine the quantities of all phases present in a 1 liter inert container at 1000 K if 10^{-6} moles of CuO and 10^{-7} moles of Cu react until the solids and vapor are in equilibrium.

	$\Delta G_f^0(1000\ K)$
CuO	-66.622 kJ mol^{-1}
Cu_2O	-77.935 kJ mol^{-1}

8. Using the data tabulated below estimate the change in C (graphite) = C (diamond) transition temperature with pressure.

	$S^0(298)$ J mol^{-1} K^{-1}	density g cm^{-3}
Diamond	2.376	3.51
Graphite	5.740	2.25

Chapter VI

Reciprocal Space and Irreducible Representations of Space Groups

VI.1 Introduction

Many aspects of the physics and chemistry of crystalline solids require the use of group theory, for example, the study of band theory, of phonon theory and infra-red and Raman active modes, of soft modes and of the Landau theory of phase transitions are all based fundamentally upon the theory of space groups and their irreducible representations. The set of pure translational symmetry operations $\{\varepsilon/\mathbf{T}_i\}$ is a subgroup of the space group of a three dimensional crystalline solid, and this leads naturally to an initial search for the irr. reps. of this subgroup. A representation is a set of matrices which are associated with the symmetry operations (in this case the pure transitions) of the group and which multiply like the operations, i.e., if ψ_1 is a matrix associated with $\mathbf{T}_1$ and ψ_2 is associated with $\mathbf{T}_2$ then $\psi_2\psi_1$ is associated with $\mathbf{T}_1 + \mathbf{T}_2$. It is not necessary that $\psi_1 \neq \psi_2$.

There is a theorem of group theory that says that if all of the operations of a group commute then the irr. reps. of the group are one dimensional i.e., the sets of basis functions contain only one function each and this function transforms into itself multiplied by a constant under the symmetry operations of the group. The pure translations do commute (vector addition is commutative) and thus the search for basis function (and irr. reps.) of the pure translational subgroup is the search for functions which change by a multiplicative constant under a translational symmetry operation.

VI.2 Reciprocal Lattice

For example, we first seek a function which transforms into itself under the translational symmetry operations i.e., a function which has the periodicity of the lattice. This function is a basis function for the "totally symmetric" representation and its representation is a string of l's, one for each $\mathbf{T}_i$.

For the purpose of finding a basis function for the totally symmetric representation of the pure translations it is useful to know a function of $\mathbf{r}$ that is incremented by an integer when $\mathbf{r}$ is incremented by $\mathbf{T}$. That is, it would help to have available a vector $\mathbf{K}$ such that

$$\mathbf{K} \cdot (\mathbf{r} + \mathbf{T}) = \mathbf{K} \cdot \mathbf{r} + \mathbf{K} \cdot \mathbf{T} \quad \text{(VI.1)}$$

$$= \mathbf{K} \cdot \mathbf{r} + \text{integer}, \quad \text{(VI.2)}$$

for with such a vector, for example

$$\sin 2\pi\mathbf{K}\cdot(\mathbf{r}+\mathbf{T}) = \sin 2\pi[\mathbf{K}\cdot\mathbf{r}+\mathbf{K}\cdot\mathbf{T}] \tag{VI.3}$$

$$= \sin 2\pi\mathbf{K}\cdot \mathrm{r} \tag{VI.4}$$

and we would have found, as desired, a function with the periodicity of the lattice. Thus we seek a $\mathbf{K}$ such that $\mathbf{K}\cdot\mathbf{T}$ = integer. We know that $\mathbf{T} = m\mathbf{a} + n\mathbf{b} + p\mathbf{c}$ and therefore $\mathbf{K} = m^*\mathbf{a}^* + n^*\mathbf{b}^* + p^*\mathbf{c}^*$ with m^*, n^*, p^* integers and $\mathbf{a}^*\cdot\mathbf{a} = 1$, $\mathbf{a}^*\cdot\mathbf{b} = \mathbf{a}^*\cdot\mathbf{c} = 0$, $\mathbf{b}^*\cdot\mathbf{b} = 1$, $\mathbf{b}^*\cdot\mathbf{a} = \mathbf{b}^*\cdot\mathbf{c} = 0$, and $\mathbf{c}^*\cdot\mathbf{c} = 1$, $\mathbf{c}^*\cdot\mathbf{b} = \mathbf{c}^*\cdot\mathbf{a} = 0$ has the desired property since, with these definitions,

$$\mathbf{K}\cdot\mathbf{T} = m^*m + n^*n + p^*p \tag{VI.5}$$

which is an integer. The definitions mean that $\mathbf{a}^*$ is perpendicular to $\mathbf{b}$ and $\mathbf{c}$, i.e., is proportional to $\mathbf{b}\times\mathbf{c}$. Taking the proportionality constant to be α,

$$\mathbf{a}^* = \alpha(\mathbf{b}\times\mathbf{c})\,, \tag{VI.6}$$

and determining the value of α by $\mathbf{a}^*\cdot\mathbf{a} = \mathbf{a}\cdot\mathbf{a}^* = 1$ yields

$$\alpha(\mathbf{a}\cdot\mathbf{b}\times\mathbf{c}) = \alpha V_{\text{cell}} = 1\,, \tag{VI.7}$$

or $\alpha = V_{\text{cel}}^{-1}$. Proceeding similarly for $\mathbf{b}^*$ and $\mathbf{c}^*$ yields

$$\mathbf{a}^* = \frac{\mathbf{b}\times\mathbf{c}}{V_{\text{cell}}}\,,\qquad \mathbf{b}^* = \frac{\mathbf{c}\times\mathbf{a}}{V_{\text{cell}}}\,,\qquad \mathbf{c} = \frac{\mathbf{a}\times\mathbf{b}}{V_{\text{cell}}}\,, \tag{VI.8}$$

and with $\mathbf{K} = m^*\mathbf{a}^* + n^*\mathbf{b}^* + p^*\mathbf{c}^*$ the functions $\sin 2\pi\mathbf{K}\cdot\mathbf{r}$, $\cos 2\pi\mathbf{K}\cdot\mathbf{r}$ and $\exp 2\pi i\mathbf{K}\cdot\mathbf{r}$ all have the periodicity of the lattice (each is a basis function for the totally symmetric irr. rep.). The definition of $\mathbf{K}$ is such that $\mathbf{K}\cdot\mathbf{r}$ is dimensionless, thus $\mathbf{K}$ has the dimension of reciprocal length and is called a reciprocal vector or, since m^*, n^* and p^* can be allowed all integral values thereby generating a lattice (a **reciprocal lattice**), $\mathbf{K}$ is called a **reciprocal lattice vector.**

The essential feature used in the definition of the reciprocal lattice vector is that $\mathbf{K}\cdot\mathbf{T}$ = integer (note: it is also common to define $\mathbf{K}$ such that $\mathbf{K}\cdot\mathbf{T}$ is an integral multiple of 2π, and this reciprocal vector is obtained from that defined here by multiplication by 2π). It follows that if $\beta\mid\mathbf{t}$ is a symmetry operation of a crystalline solid then $\mathbf{K}\cdot\beta\mathbf{T}$ is an integer if $\mathbf{K}\cdot\mathbf{T}$ is. However if $\mathbf{K}\cdot\beta\mathbf{T}$ is an integer then so too is $\beta^{-1}\mathbf{K}\cdot\mathbf{T}$. This is true for a given $\mathbf{T}$ and all of the reciprocal vectors generated by the rotational part of the symmetry operations of the space group. Thus if $\mathbf{K}$ is a reciprocal lattice vector so too is $\beta\mathbf{K}$ and the reciprocal lattice has the same rotational symmetry as the real lattice.

VI.3 Reciprocal Space

It becomes necessary when one wishes to consider periodicities different from that of the lattice, i.e., to consider basis functions for irr. reps. of the translational subgroup other than the totally symmetric, to generalize the reciprocal vector concept to include vectors with nonintegral components of **a***, **b*** and/or **c***. Therefore the **wave vector, k**, which is a general, nonintegral component vector in **reciprocal space** is introduced. For example, μ**a*** corresponds to a wave vector along the **a*** direction.

We have shown that among the rotational symmetry operations of the reciprocal lattice are included those which make up the crystal class of the space group. These rotational symmetry operations carry wave vectors into symmetrically equivalent wave vectors and thus are symmetry operations of reciprocal space. The general wave vectors, under the β operations of the crystal class, fall into three cases. Either:

1. they transform into different wave vectors which do not differ from the original by a reciprocal lattice vector, or
2. they transform into different wave vectors which do differ from the original by a reciprocal lattice vector, or
3. they transform into themselves.

For example, μ**a*** with $0 < \mu < \frac{1}{2}$ transforms into itself under C_2 or σ_z (case 3) but into $-\mu$**a*** under i and C_2 (case 1). On the other hand, $-\frac{1}{2}$ **a*** transforms into $-\frac{1}{2}$ **a*** under i and C_2 and thus **a***/2 falls under case 2 for these operations.

Cases 2 and 3 can be included in the same category if it is specified that symmetry operations in this category carry **K** into itself modulo a reciprocal lattice vector, i.e., **k** into **k** + **K** including **K** = 0. The set of all β parts of the symmetry operations that carry **k** into **k** modulo **K** form a point group called the **point group of the wave vector,** g_0(**k**). This statement is proven below. The remaining β parts of the space group operations carry the **k** vector into

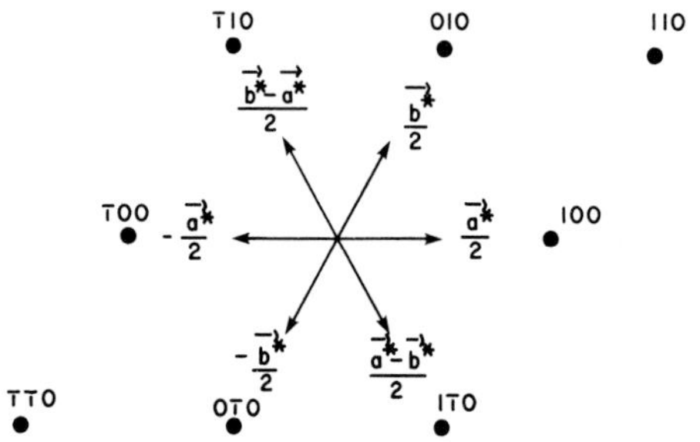

Fig. VI.1

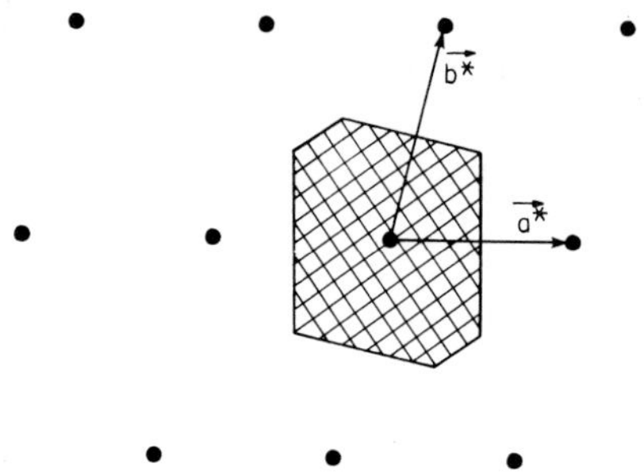

Fig. VI.2

Fig. VI.1. Reciprocal space for a hexagonal lattice showing vectors similar to **a***/2 and in the star of **a***/2.

Fig. VI.2. A monoclinic reciprocal lattice projected along the unique axis showing the projected first Brillouin zone. The zone extends to $\pm$ **c***/2 relative to the **a*** — **b*** plane.

other **k** vectors; those which do not differ by a reciprocal lattice vector are said to form a **star**.

For example, consider the wave vector **a***/2 in the case of a hexagonal space group in the class D_{6h} (Fig. VI.1). The vector that result from the β operations on **a***/2 are ±**a***/2, ±**b***/2 and ±(**a*** − **b***)/2. Of these **a***/2, **b***/2 and (**a*** − **b***)/2 from a star. Those symmetry operations that carry **a***/2 into ±**a***/2 (into **a***/2 modulo **a***) are ε, C_2, C_{2y}, $C_{2(2x+y)}$, i, σ_z, σ_y and σ_{2x+y}, i.e., the symmetry operations of D_{2h} expressed in hexagonal coordinates. Thus the point group of the wave vector **a***/2 is D_{2h}. The point groups of the other wave vectors in the star **b***/2 and (**a*** − **b***)/2, are also D_{2h}, but these point groups consist of different sets of two-fold rotations and reflections than those appropriate to **a***/2 (and, of course, these two sets also differ from each other).

To prove that the set of β operations of the crystal class that carry **k** into **k** modulo **K** form a group in general we must prove closure of the set, that the set contains all inverses and an identity and that the binary operation is associative. First, regarding closure, let

$$\beta_1 \mathbf{k} = \mathbf{k} + \mathbf{K}_1 \tag{VI.9}$$

and

$$\beta_2 \mathbf{k} = \mathbf{k} + \mathbf{K}_2 \,. \tag{VI.10}$$

Then

$$\beta_1 \beta_2 \mathbf{k} = \beta_1 \mathbf{k} + \beta_1 \mathbf{K}_2 \tag{VI.11}$$

$$= \mathbf{k} + \mathbf{K}_1 + \beta_1 \mathbf{K}_2 \tag{VI.12}$$

and from the fact that $\beta_1 \mathbf{K}_2$ is necessarily a reciprocal lattice vector, and thus $\mathbf{K}_1 + \beta \mathbf{K}_2$ is, we can write

$$\beta_1 \beta_2 \mathbf{k} = \mathbf{k} + \mathbf{K}_3 \tag{VI.13}$$

where $\mathbf{K}_3 = \mathbf{K}_1 + \beta \mathbf{K}_2$ and closure is proven. Next consider the existence of an identity;

$$\varepsilon \mathbf{k} = \mathbf{k} + \mathbf{K} \tag{VI.14}$$

with $\mathbf{K} = 0$ and thus ε is in the set. Regarding inverses, if

$$\beta \mathbf{k} = \mathbf{k} + \mathbf{K} \tag{VI.15}$$

then

$$\beta^{-1} \beta \mathbf{k} = \mathbf{k} = \beta^{-1} \mathbf{k} + \beta^{-1} \mathbf{K} \tag{VI.16}$$

or

$$\beta^{-1} \mathbf{k} = \mathbf{k} - \beta^{-1} \mathbf{K} \tag{VI.17}$$

Table VI.1. The Multiplication Table for a Loaded Rep. of $P2_1/m$ at $\mathbf{c}^*/2$.

$\beta_2 \downarrow$ $\beta_1 \rightarrow$	$\tilde{\Gamma}(\varepsilon)$	$\tilde{\Gamma}(C_{2z})$	$\tilde{\Gamma}(i)$	$\tilde{\Gamma}(\sigma_z)$
$\tilde{\Gamma}(\varepsilon)$	$\tilde{\Gamma}(\varepsilon)$	$\tilde{\Gamma}(C_{2z})$	$\tilde{\Gamma}(i)$	$\tilde{\Gamma}(\sigma_z)$
$\tilde{\Gamma}(C_{2z})$	$\tilde{\Gamma}(C_{2z})$	$\tilde{\Gamma}(\varepsilon)$	$\tilde{\Gamma}(\sigma_z)$	$\tilde{\Gamma}(i)$
$\tilde{\Gamma}(i)$	$\tilde{\Gamma}(i)$	$-\tilde{\Gamma}(\sigma_z)$	$\tilde{\Gamma}(\varepsilon)$	$-\tilde{\Gamma}(C_{2z})$
$\tilde{\Gamma}(\sigma_z)$	$\tilde{\Gamma}(\sigma_z)$	$-\tilde{\Gamma}(i)$	$\tilde{\Gamma}(C_{2z})$	$-\tilde{\Gamma}(\varepsilon)$

and since $-\beta^{-1}\mathbf{K}$ is necessarily a reciprocal lattice vector we can set $\mathbf{K}' = -\beta^{-1}\mathbf{K}$, and

$$\beta^{-1}\mathbf{k} = \mathbf{k} + \mathbf{K}' \tag{VI.18}$$

showing that all β's have inverses within the set. Associativity is a property of the binary combination operation in the crystal class and thus this operation is associative. Hence the set of all operations of the crystal class that carry $\mathbf{k}$ into $\mathbf{k}$ modulo $\mathbf{K}$ form a group $g_0(\mathbf{k})$.

The reason for introducing the concept of reciprocal space is that it provides access to basis functions and representations of the translational symmetry operations. So far as the translational subgroup is concerned, $\exp(2\pi i\mathbf{k} \cdot \mathrm{r})$ is a basis function for the translational symmetry operations, i.e., transforming $\mathbf{r}$ into $\mathbf{r} + \mathbf{T}$ transforms $\exp(2\pi i\mathbf{k} \cdot \mathbf{r})$ into $\exp(2\pi i\mathbf{k} \cdot (\mathbf{r} + \mathbf{T})) = \exp(2\pi i\mathbf{k} \cdot \mathbf{T}) \exp(2\pi i\mathbf{k} \cdot \mathbf{r})$ and thus into a constant $(\exp(2\pi i\mathbf{k} \cdot \mathbf{T}))$ times itself.

For example if $\mathbf{k} = \mathbf{a}^*/2$ then $\exp 2\pi i\mathbf{k} \cdot \mathbf{r} = \exp(\pi i x)$ and this function is transformed into $\exp(\pi i(x + m)) = \exp(m\pi i) \exp(\pi i x)$ by the pure translation $\mathbf{T} = m\mathbf{a} + n\mathbf{b} + p\mathbf{c}$, and is thus symmetric with respect to translations with m even $(\exp(m\pi i) = 1)$ and antisymmetric with respect to translations with m odd $(\exp(m\pi i) = -1)$.

If we consider, on the other hand, transforming $\mathbf{k}$ into $\mathbf{k} + \mathbf{K}$ then $\exp(2\pi i\mathbf{k} \cdot \mathbf{r})$ transforms into $\exp(2\pi i(\mathbf{k} + \mathbf{K}) \cdot \mathbf{r}) = \exp(2\pi i\mathbf{K} \cdot \mathbf{r}) \exp(2\pi i\mathbf{k} \cdot \mathbf{r})$, i.e., into the product of the function with a totally symmetric function. So far as characterizing functions by symmetry is concerned, these functions are basis functions for the same irr. rep. That is, for example, both $\exp((2\pi i\mathbf{a}^*/2) \cdot \mathbf{r}) = \exp(\pi i x)$ and $\exp(2\pi i(\mathbf{a}^*/2 + \mathbf{K}) \cdot \mathbf{r}) = \exp(2\pi i\mathbf{K} \cdot \mathbf{r}) \exp(\pi i x)$ are basis functions for the irr. rep. of the translational subgroup that is symmetric with respect to even $\mathbf{a}$ and antisymmetric with respect to odd $\mathbf{a}$ translations.

The above means that functions of all possible translational symmetries are considered when one considers the set of all wave vectors defined so that $\mathbf{k}$ modulo $\mathbf{K}$ does not belong to the set. This fact leads to the definition of the Brillouin zone. The first Brillouin zone is the volume of reciprocal space containing all wave vectors from the origin (called point Γ) such that $\mathbf{k}$ modulo $\mathbf{K}$ does not belong to the set plus (in order to get a closed set) all vectors terminating on the boundary of the set (even though the vectors terminating on the boundary may differ by **K**). In order to construct this zone it is only necessary to enclose a volume about a reciprocal lattice point (taken as point Γ) by constructing planes which are perpendicular bisectors of all lines connecting lattice points to the origin as shown for a monoclinic reciprocal lattice in projection in Fig. VI.2. Note that $\pm\mathbf{a}^*/2$, $\pm\mathbf{b}^*/2$ and $\pm(\mathbf{a}^* - \mathbf{b}^*)/2$ all terminate on the zone boundary and are therefore included within the zone even though the members of each pair differ by a reciprocal lattice vector.

This fact gives rise to special symmetries for some points on the boundary of the zone. For example, in the monoclinic case (Fig. VI.2), assuming the crystal class to be C_{2h}, $\mu\mathbf{a}^*$ with $0 < \mu < \frac{1}{2}$ is carried into itself under σ_z, but not i and C_{2z}, and thus $\{\varepsilon, \sigma_z\}$ constitutes $g_0(\mu\mathbf{a}^*)$. On the other hand, when $\mu = \frac{1}{2}$ both i and C_2 carry $\mathbf{a}^*/2$ into $-\mathbf{a}^*/2$, which differs from $\mathbf{a}^*/2$ by $\mathbf{a}^*$, and thus both C_{2z} and i are in the point group of the wave vector $\mathbf{a}^*/2$, i.e., $g_0(\mathbf{a}^*/2)$ is C_{2h}.

VI.4 Irreducible Representation of Space Groups

The irr. reps. of space groups are, for a large number of cases, simply related to the irr. reps. of the point groups $g_0(\mathbf{k})$. In order to discover these cases and find the irr. reps. consider as a possible irr. rep. for $\beta_i \mid \mathbf{t}_i$ at point **k** in the first Brillouin zone

$$\Gamma(\beta_i \mid \mathbf{t}_i) = (\exp(2\pi i\mathbf{k}\cdot\mathbf{t}))\,\Gamma(\beta_i) \tag{VI.19}$$

where $\Gamma(\beta)$ is the irr. rep. of β_i in $g_0(\mathbf{k})$, and as has been shown, $\exp 2\pi i\mathbf{k}\cdot\mathbf{t}$, is the irr. rep. of $\mathbf{t}_i$ when $\mathbf{t}_i$ is in the group of the pure translations (which is not always the case). It is because of this last parenthetical remark that the trial irr. rep. might not work. If such a form is to be a representation then

$$\Gamma(\beta_2 \mid \mathbf{t}_2)\,\Gamma(\beta_1 \mid \mathbf{t}_1) = \Gamma(\beta_1\beta_2 \mid \beta_2\mathbf{t}_1 + \mathbf{t}_2) \tag{VI.20}$$

i.e.,

$$\exp(2\pi i\mathbf{k}\cdot\mathbf{t}_2)\,\Gamma(\beta_2)\exp(2\pi i\mathbf{k}\cdot\mathbf{t}_1)\,\Gamma(\beta_1) = \exp(2\pi i\mathbf{k}\cdot(\beta_2\mathbf{t}_1 + \mathbf{t}_2))\,\Gamma(\beta_2\beta_1). \tag{VI.21}$$

Since $\Gamma(\beta_2)\,\Gamma(\beta_1) = \Gamma(\beta_2\beta_1)$, and since $\exp(2\pi i\mathbf{k}\cdot\mathbf{t}_2)$ can be cancelled from both sides, the equality holds if and only if

$$\mathbf{k}\cdot\mathbf{t}_1 = \mathbf{k}\cdot\beta_2\mathbf{t}_1 \tag{VI.22}$$

for all space group operations with rotational parts in $g_0(\mathbf{k})$. This equality can be seen to be valid for a number of cases:

1. $\mathbf{k} = 0$, i.e., at point Γ for all space groups,

2. $\mathbf{t}_i = \mathbf{T}_i$ for all operations with rotational parts in $g_0(\mathbf{k})$ e.g., at all points of reciprocal space for symmorphic space groups,
3. when $\beta\mathbf{t}_1 = \mathbf{t}_1$ for all β in $g_0(\mathbf{k})$ e.g., when $g_0(\mathbf{k}) = \varepsilon$ (general points of all space groups) and when $\mathbf{t}_i$ lies along an axis or in a plane and there are no intersecting axes or planes,
4. when $\mathbf{k}$ is perpendicular to $\mathbf{t}$ and $\beta\mathbf{t}$ e.g., when $\mathbf{k} = \mathbf{a}^*/2$ and $\mathbf{t} = \mathbf{c}/2$.

These cases cover most of the points in most Brillouin zones. The cases that are not covered are those, for example, for which $\mathbf{k}$ and $\mathbf{t}$ are parallel (e.g., $\mathbf{k} = \mathbf{c}^*/2$ and $\mathbf{t} = \mathbf{c}/2$) for a nonsymmorphic space group such as $P2_1/m$. In cases such as these the irr. reps. suggested above need not work and, in fact, as well shall show, do not work. Therefore a different approach to obtaining irr. reps. at these points is needed.

An approach to this problem, motivated by the above, is to define what is called a "loaded" representation (which is, in fact, not necessarily a representation at all) for β_1 in $g_0(\mathbf{k})$ according to

$$\tilde{\Gamma}(\beta) = \exp(-2\pi i\mathbf{k}\cdot\mathbf{t}_i)\,\Gamma(\beta\,|\,\mathbf{t}) \tag{VI.23}$$

and we note that $\tilde{\Gamma}(\beta_i)$ is in fact $\Gamma(\beta_i)$ in the cases 1–4 listed above. A multiplication table for the $\tilde{\Gamma}$'s can be generated, for since $\Gamma(\beta_i\,|\,\mathbf{t}_i)$ must multiply like the symmetry operations.

$$\tilde{\Gamma}(\beta_2)\,\tilde{\Gamma}(\beta_1) = \exp(-2\pi i\mathbf{k}\cdot(\mathbf{t}_2 + \mathbf{t}_1))\,\Gamma(\beta_2\,|\,\mathbf{t}_2)\,\Gamma(\beta_1\,|\,\mathbf{t}_1) \tag{VI.24}$$

$$= \exp(-2\pi i\mathbf{k}\cdot(\mathbf{t}_2 + \mathbf{t}_1))\,\Gamma(\beta_2\beta_1\,|\,\beta_2\mathbf{t}_1 + \mathbf{t}_2) \tag{VI.25}$$

$$= \exp(2\pi i\mathbf{k}\cdot(\beta_2\mathbf{t}_1 - \mathbf{t}_1))\,\tilde{\Gamma}(\beta_2\beta_1)\,. \tag{VI.26}$$

For example in the case of $P2_1/m$ at $\mathbf{c}^*/2$ the essential symmetry operations are $\varepsilon\,|\,0$, $C_{2z}\,|\,\mathbf{c}/2$, $i\,|\,0$, $\sigma_z\,|\,\mathbf{c}/2$. Adopting the convention that the symmetry operations listed along the top of a table correspond to those operations performed first ($\beta_1\,|\,\mathbf{t}_1$ in the above) and those listed down the side to those performed second ($\beta_2\,|\,\mathbf{t}_2$), the multiplication table (Table VI.1) for the loaded reps. can be constructed. Realizing that $\tilde{\Gamma}(\beta_2)\,\tilde{\Gamma}(\beta_1) = \tilde{\Gamma}(\beta_2\beta_1)$ when $t_1 = 0$ allows all of the entries in the ε and i columns to be directly introduced as $\tilde{\Gamma}(\beta_2\beta_1)$, the same is true when β_2 carries t_1 into itself, i.e., for the rows led by ε and C_{2z}. This leaves four entries, $\tilde{\Gamma}(i)\,\tilde{\Gamma}(\sigma_z)$, $\tilde{\Gamma}(i)\tilde{\Gamma}(C_{2z})$, $\tilde{\Gamma}(\sigma_z)\,\tilde{\Gamma}(C_{2z})$ and $\tilde{\Gamma}(\sigma_z)\tilde{\Gamma}(\sigma_z)$ to be determined. In all of these cases $\beta_2\mathbf{t}_1 = -\mathbf{t}_1(= -\mathbf{c}/2)$ and thus

$$\exp(2\pi i\mathbf{k}\cdot(\beta_2 t_1 - t_1)) = \exp\left(2\pi i\,\frac{\mathbf{c}^*}{2}\cdot\left(-\frac{\mathbf{c}}{2} - \frac{\mathbf{c}}{2}\right)\right) = \exp(-\pi i) = -1\,, \tag{VI.27}$$

and thus $\tilde{\Gamma}(\beta_2)\,\tilde{\Gamma}(\beta_1) = -\tilde{\Gamma}(\beta_2\beta_1)$ in these cases. The completed table is as shown (Table VI.1) and it presents two noteworthy features. First of these is that not all of the loaded reps. commute, e.g.,

$$\tilde{\Gamma}(i)\,\tilde{\Gamma}(\sigma\ ) = -\tilde{\Gamma}(\sigma\ )\,\tilde{\Gamma}(i) \tag{VI.28}$$

and thus the $\tilde{\Gamma}$'s can not be numbers (real or complex), but rather must be at least 2×2 matrices. The second feature of note is that no inverse for $\tilde{\Gamma}(\sigma_z)$ is present in the table, i.e., the $\tilde{\Gamma}$'s are not a representation of a group. Nonetheless the $\tilde{\Gamma}$'s are useful, as shown in what follows.

Since we know that there is no one-dimensional solution to the multiplication table we seek the solution in 2×2 matrices. We set $\tilde{\Gamma}(\varepsilon) = \begin{pmatrix} 1 & 0 \\ 0 & 1 \end{pmatrix}$ and $\tilde{\Gamma}(C_{2z}) = \begin{pmatrix} a & b \\ c & d \end{pmatrix}$

and from $\tilde{\Gamma}(C_{2z})\,\tilde{\Gamma}(C_{2z}) = -\tilde{\Gamma}(\varepsilon)$ we find that

$$a^2 + bc = d^2 + bc = -1 \tag{VI.29}$$

and

$$b(a + d) = c(a + d) = 0\,. \tag{VI.30}$$

A solution is $b = -c = 1$ and $a = d = 0$, Letting $\tilde{\Gamma}(i) = \begin{pmatrix} e & f \\ g & h \end{pmatrix}$ and using

$$\begin{pmatrix} 0 & 1 \\ -1 & 0 \end{pmatrix}\begin{pmatrix} e & f \\ g & h \end{pmatrix} = -\begin{pmatrix} e & f \\ g & h \end{pmatrix}\begin{pmatrix} 0 & 1 \\ -1 & 0 \end{pmatrix}, \tag{VI.31}$$

where $\begin{pmatrix} 0 & 1 \\ -1 & 0 \end{pmatrix}$ is the matrix found for $\tilde{\Gamma}(C_{2z})$, and

$$\begin{pmatrix} e & f \\ g & h \end{pmatrix}\begin{pmatrix} e & f \\ g & h \end{pmatrix} = \begin{pmatrix} 1 & 0 \\ 0 & 1 \end{pmatrix} \tag{VI.32}$$

yields, from the former $g = f$ and $-h = e$ and from the latter

$$e^2 + fg = h^2 + fg = 1 \tag{VI.33}$$

and

$$f(e + h) = g(e + h) = 0\,, \tag{VI.34}$$

and thus a solution is $f = g = 1$ or $\tilde{\Gamma}(i) = \begin{pmatrix} 0 & 1 \\ 1 & 0 \end{pmatrix}$. It then follows that

$$\tilde{\Gamma}(C_{2z}) = \begin{pmatrix} 1 & 0 \\ 0 & -1 \end{pmatrix}, \tag{VI.35}$$

and thus we have

β	ε	C_{2z}	i	σ_z
$\tilde{\Gamma}(\beta)$	$\begin{pmatrix} 1 & 0 \\ 0 & 1 \end{pmatrix}$	$\begin{pmatrix} 1 & 0 \\ 0 & -1 \end{pmatrix}$	$\begin{pmatrix} 0 & 1 \\ 1 & 0 \end{pmatrix}$	$\begin{pmatrix} 0 & 1 \\ -1 & 0 \end{pmatrix}$

.

Using $\Gamma(\beta \mid \mathbf{t}) = \exp(2\pi i \mathbf{k} \cdot \mathbf{t})\,\tilde{\Gamma}(\beta)$ we obtain the small representation of $P2_1/m$ at $\mathbf{c}^*/2$:

$\beta \mid \mathbf{t}$	$\varepsilon \mid 000$	$C_{2z} \mid 00\frac{1}{2}$	$i \mid 000$	$\sigma_z \mid 00\frac{1}{2}$
$\Gamma(\beta \mid \mathbf{t})$	$\begin{pmatrix}1&0\\0&1\end{pmatrix}$	$\begin{pmatrix}i&0\\0&-i\end{pmatrix}$	$\begin{pmatrix}0&1\\1&0\end{pmatrix}$	$\begin{pmatrix}0&i\\-i&0\end{pmatrix}$

from which the irr. rep. of the space group can be generated. Note that the functions $\exp \pi iz = \varphi_1$ and $\exp -\pi iz = \varphi_2$ are basis functions for this irr. rep. (Table VI.2).

Table VI.2.

$\varepsilon \mid 000$	x, y, z	$\begin{pmatrix}1&0\\0&1\end{pmatrix}\begin{pmatrix}e^{\pi iz}\\e^{-\pi iz}\end{pmatrix} = \begin{pmatrix}e^{\pi iz}\\e^{-\pi iz}\end{pmatrix}$
$C_{2z} \mid 00\frac{1}{2}$	$\bar{x}, \bar{y}, z + \frac{1}{2}$	$\begin{pmatrix}i&0\\0&-i\end{pmatrix}\begin{pmatrix}e^{\pi iz}\\e^{-\pi iz}\end{pmatrix} = \begin{pmatrix}e^{\pi i\left(z+\frac{1}{2}\right)}\\e^{-\pi i\left(z+\frac{1}{2}\right)}\end{pmatrix}$
$i \mid 000$	$\bar{x}, \bar{y}, \bar{z}$	$\begin{pmatrix}0&1\\1&0\end{pmatrix}\begin{pmatrix}e^{\pi iz}\\e^{-\pi iz}\end{pmatrix} = \begin{pmatrix}e^{-\pi iz}\\e^{\pi iz}\end{pmatrix}$
$\sigma_z \mid 00\frac{1}{2}$	$x, y, \bar{z} + \frac{1}{2}$	$\begin{pmatrix}0&i\\-i&0\end{pmatrix}\begin{pmatrix}e^{\pi iz}\\e^{-\pi iz}\end{pmatrix} = \begin{pmatrix}e^{\pi i\left(\bar{z}+\frac{1}{2}\right)}\\e^{-\pi i\left(\bar{z}+\frac{1}{2}\right)}\end{pmatrix}$

We are thus able to straightforwardly obtain space group irr. reps. from those of $g_0(\mathbf{k})$ in a great many cases, and to find loaded irr. reps. from which those of the space group can be inferred in others. The irr. reps. have the physical significance that properties or changes that correspond to a function of position in the crystal that is a basis function, or a linear combination of basis functions, for a given irr. rep. are related by symmetry to other properties or changes which correspond to other combinations of those basis functions, but are unrelated to those properties or changes which correspond to functions which are combinations of basis functions for other irr. reps. This fact has been exploited by Landau in the theory concerning symmetry aspects of second-order phase transitions that bears his name and is the major topic of Chap. VII.

VI.5 Problems

1. Plot the $\mathbf{a}^* - \mathbf{b}^*$ plane of a reciprocal lattice for a hexagonal cell. Find the volume of the reciprocal unit cell in terms of a, b, c.

2. Show that the reciprocal lattice of a face centered cubic lattice is body centered cubic.
3. Find the point group of the wave vectors **a***/2, **b***/2, (**a*** + **b***)/2 for
   a. $P2_1/m$.
   b. $P6_3/mmc$.
4. Show that the irr. reps. of 3a at **a***/2 are small irr. reps. of the space group but those at **b***/2 are not.
5. Find the irr. rep. of $P2_1/m$ at **a***/2.
6. Find the irr. rep. of P_4 at **a***/2.
7. Sketch the Brillouin zone for the face-centered cubic lattice.
8. Find the irr. rep. of P2/c at **c***/2.

Chapter VII

Second-Order Phase Transitions

VII.1 Introduction

This chapter treats the thermodynamic and group theoretical techniques useful in the consideration of phase transitions which occur without the coexistence of two phases, i.e., without the nucleation and growth of a new phase. The transitions under consideration occur with a change of symmetry at a certain thermodynamic state during a continuous structure change through that state. Such crystal-structure changes may be of three types: order-disorder, displacive and a combination of order-disorder and displacive.

The classical example of a pure order-disorder transition that occurs with a continuous change in structure is that in CuZn which is called the β-β′ brass transition [34]. At very low temperatures there is a tendency for CuZn to adopt the CsCl structure (Fig. IV.5), i.e., for the Cu atoms to preferentially occupy a simple cubic lattice with origin at 0, 0, 0 and for the Zn atoms to occupy a similar lattice with origin at $\frac{1}{2}, \frac{1}{2}, \frac{1}{2}$ (or, what is equivalent by a change of origin of the structure, vice versa). As the temperature is increased there is a tendency for the atoms to interchange at random, an interchange which increases the configuration entropy of the system ($\Delta S = R \ln 2$ for the change from CsCl-type to completely random distribution of Cu and Zn atoms on a bcc lattice).

An order-disorder parameter, η, can be defined

$$\eta = 2f_{Cu} - 1 \qquad \text{(VII.1)}$$

where f_{Cu} is the fraction of 0, 0, 0 sites occupied by Cu atoms. This parameter is 0 for the random distribution and 1 for the completely ordered CsCl-type structure. From the point of view of symmetry there are symmetry elements present in the $\eta = 0$ case that are missing when $\eta \neq 0$, for example translation by $\frac{1}{2}, \frac{1}{2}, \frac{1}{2}$ (i.e., $(\mathbf{a} + \mathbf{b} + \mathbf{c})/2$) is a symmetry operation only when $\eta = 0$. It is experimentally observed that η does vanish continuously at sufficiently high temperatures and is, as stated above, nonzero at low temperatures. Since a symmetry operation can be lost (or gained) only at a state point it follows that a phase transition can be defined as a symmetry change, and a transition point (e.g., temperature) as the state which seperates $\eta = 0$ from $\eta \neq 0$ structures (provided the order-disorder process takes place continuously, as apparently is the case).

Some other examples of apparently continuous order-disorder processes are the ordering of vacancies in the Sc substructure of $Sc_{1-x}S$ in alternate metal planes along the (1, 1, 1) direction of the defect NaCl-type structure (a change from Fm3m to

$R\bar{3}m$ symmetry [35]) and the ordering of Cr vacancies in defect $Cr_{1-x}S$ in alternate planes along the c-axis of the defect hexagonal ($P6_3/mmc$)NiAs-type to yield the defect CdI_2-type ($P\bar{6}m2$) structure, [36]. These order-disorder transitions are discussed in Chap. IX.

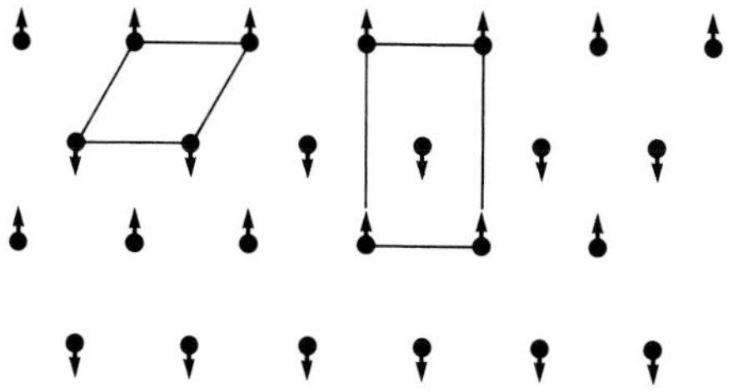

Fig. VII.1. Metal atom positions in the z = 0 metal atom layer of the NiAs-type structure with the distortion to MnP-type observed for VS indicated by the arrows. The metal atoms at $z = \frac{1}{2}$ move in the opposite direction. The hexagonal (NiAs) and orthorhombic (MnP) cells are outlined.

An example of a displacive transition which is believed to occur continuously is the NiAs-type to MnP-type phase transition in CoAs [37] and VS [38]. In this case the atoms are observed to move in a concerted fashion such as to destroy the hexagonal symmetry ($P6_3/mmc$) but maintain orthorhombic symmetry (Pcmn). The movement of the metal atoms in the hexagonal layer at $z = 0$ is shown schematically in Fig. VII.1. In this case η can be defined as the length of the vectors describing the distortion, and with normalization this η varies between 1 at 0 K and 0 at the transition temperature.

The third type of continuous phase transition combines displacive and order-disorder changes and therefore leads necessarily to an incommensurate ordering as has recently been described [39].

VII.2 Thermodynamics of Second-Order Phase Transitions

At first, in order to appreciate the nature of the phase transitions under discussion, it is desirable to understand the thermodynamic consequences of the proposed phenomena. They have in common the following description: in some region of thermodynamic space the sample has one set of symmetry elements and in another region a different set, however efforts to discover states at which samples with the two sets coexist in equilibrium fail and it is concluded that the transition occurs without this coexistence. A phase diagram in this case, with one symmetry labeled α and the other β, and with the variables T and X (mole fraction) is shown in Fig. VII.2a.

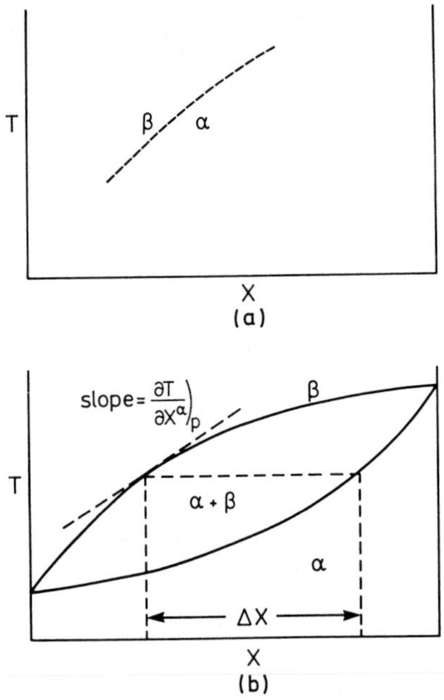

Fig. VII.2 a, b. Schematic T-x phase diagrams. **a** continuous structure change **b** two phases coexist in equilibrium.

The Gibbs-Konovalow (G-K) equation

$$\left(\frac{\partial T}{\partial X_A^\alpha}\right)_P = -\frac{\left\{\left(\frac{\partial \mu_A^\alpha}{\partial X_A^\alpha}\right)_{T,P} + \left(\frac{\partial \mu_B^\alpha}{\partial X_B^\alpha}\right)_{T,P}\right\} \Delta X}{\Delta \bar{S} + (\bar{S}_A^\alpha - \bar{S}_B^\alpha)\, \Delta X} \tag{VII.2}$$

where $\left(\frac{\partial T}{\partial X_A^\alpha}\right)_P$ is the slope of a phase boundary on a T-X diagram (see Fig. VII.2b), $\left(\frac{\partial \mu_A^\alpha}{\partial X_A^\alpha}\right)_{T,P}$ is the mole fraction partial derivative of the chemical potential in a phase (α) bounded by the phase boundary, ΔX is the difference in the mole fraction of A in the two phases separated by the two-phase region enclosed by the boundary and $\Delta \bar{S}$ and $\bar{S}_A^\alpha$ are the difference in molar entropy of the phases and the partial molar entropy of A in α, respectively, is the differential equation of the lines that appear on binary T-X phase diagrams and it provides the general theoretical description of such lines.

Applying this equation to the case in question we must first note that $\left(\frac{\partial \mu}{\partial X}\right)_{T,P}$ is not infinite (for ideal solutions it is RT/X) except at the $X = 0$ boundaries of phase diagrams. Thus for changes in symmetry occuring with continuous changes in structure in nonstoichiometric compounds, the fact that $\Delta X = 0$ and $\left(\frac{\partial T}{\partial X_A^\alpha}\right)_P \neq 0$ (see Fig. VII.2) requires that $\Delta S = 0$, i.e., that the entropy changes continuously at the transition point. Note also that the continuity of the process means also that $\Delta V = 0$ (the volume changes continuously). Processes for which ΔS and ΔV are zero are called second-order phase transitions.

If an attempt is made to plot G vs T as in Fig. VII.3 for the case of $\Delta S = 0$ a contradiction appears. Because $\Delta S = 0$ means that the G^{α} and G^{β} curves are tangent at T (see Fig. VII.4) and hence $G^{\alpha} < G^{\beta}$ for both $T > T_t$ and $T < T_t$ it would appear that there is no symmetry change. The contradiction is eliminated when it is recognized that the α symmetry cannot exist for $T > T_t$ ($T < T_t$ is discussed in a later section). As the temperature increases from $T = 0$ to $T = T_t$ η approaches zero and vanishes at T_t. The physical meaning of a continuation of η to negative values is an ordering or distortion which inverts that for $\eta > 0$. For example, the continuation of η to negative values in the CuZn case would have the meaning of ordering Zn atoms at the original Cu positions and in the NiAs-type to MnP-type case would have the meaning of inverting the arrows of Fig. VII.1. In either case continuation to $\eta < 0$ would correspond to a return to the structure stable at $T < T_t$, but with a change of origin. It follows that the expansion of G in T on both sides of T_t was not meaningful for symmetry α — only the side of T_t corresponding to $\eta > 0$ is physically meaningful.

There remains another distinction that can be made between the G vs. T diagrams appropriate to first- and second-order phase transitions. This has to do with the area between the α and β curves which, in the second-order case, is a region that is in principle accessible to the system. The β curve for $T < T_t$ corresponds to metastable $\eta = 0$, a possible physical state in the order-disorder case because there is an energy barrier to diffusion, but only theoretically accessible in the displacive case because there is no barrier to distortion. However the choice of the $\eta = 0$ curve for continuation to $T < T_t$ is arbitrary — any $\eta \neq 0$ curve could also be continued. Said in a different way, because of the continuity of the process there is a continuity of states accessible to the system corrsponding to the area between the α and β curves. This is not the case when the transition is

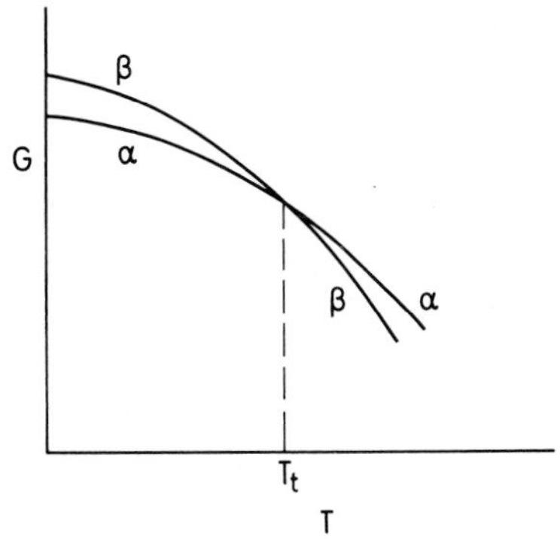

Fig. VII.3

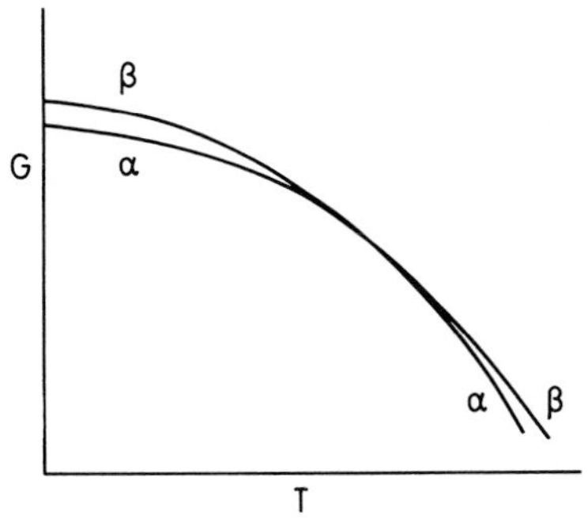

Fig. VII.4

Fig. VII.3. Schematic G vs. T (at constant P, X) phase diagram. A first-order phase transition occurs at $T = T_t$.

Fig. VII.4. G vs T for the second-order case with an unresolved problem (see text).

first-order (Fig. VII.3). The G vs. T diagram appropriate to the above considerations is shown in Fig. VII.5.

In summary, a second-order phase transition is a symmetry change which occurs with $\Delta S = \Delta V = 0$ at the transition point. It occurs during a continuous change in structure of the displacive or order-disorder type. The symmetry change must occur at a point since it is not possible for a structure to have a fraction of a symmetry operation. The point of loss or gain of symmetry is the point at which η goes to zero and it is the point at which the G vs T area (Fig. VII.5) converges to a line (volume becomes a surface in G-T-P space, etc.). At this point α and β do not coexist, but rather they become indistinguishable. At this point there is a discontinuity in C_p which will be discussed further in the next section.

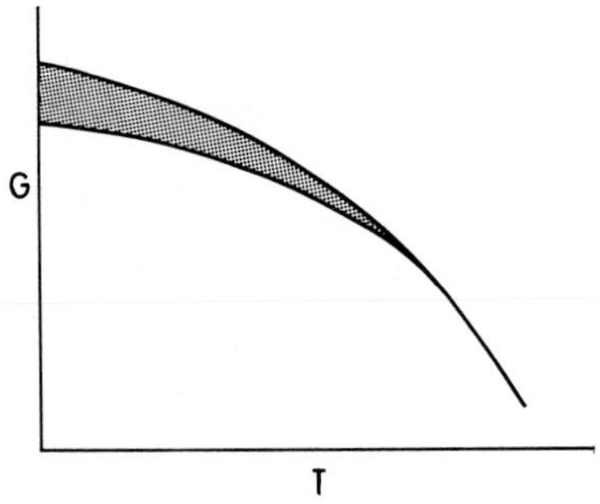

Fig. VII.5

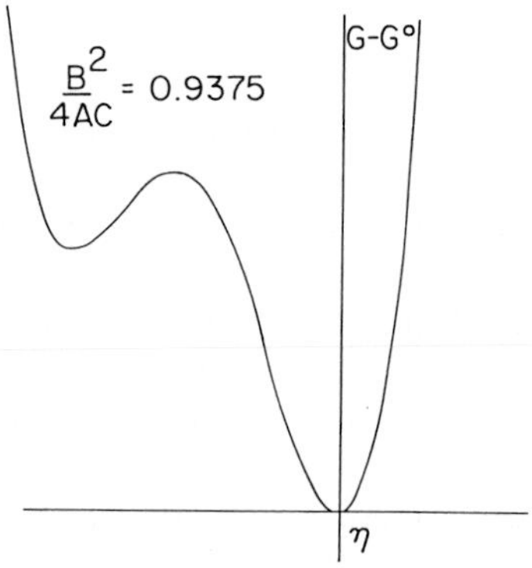

Fig. VII.6

Fig. VII.5. G vs T for a second-order phase transition. The low-temperature phase is assumed to be the phase with lower symmetry.
Fig. VII.6. G—G° vs. η (A,B,C > 0).

It is perhaps worthwhile to note that there undoubtedly exist cases of first-order phase transitions which appear to be second-order. Such cases arise when the two G vs T curves of Fig. VII.3 correspond to nearly the same structure in the neighborhood of T_t. A fundamental distinction between first- and second-order processes, namely the coexistence of two phases in equilibrium in the first-order case, is then not useful because fluctuations in the experimental variables in either space or time will obscure the distinction. The result is that while it is possible to state that observed coexistence at equilibrium implies a first-order process, the apparent abscence of such coexistence does not require a second-order process unless it can be determined that there is no coexistence down to the level of statistical fluctuations. Furthermore, it is not possible to say, simply because η decreases extremely rapidly to zero at T_t that a process is first-order because such rapid changes are allowed in a second-order process. It is not possible to devise an experiment which differentiates between an arbitrarily steep slope and a discontinuity. Therefore in cases where coexistence at equilibrium cannot be demonstrated we are in need of additional criteria for the determination of the order of a transition and we turn to the theory of Landau [34].

VII.3 Landau Theory (Without Symmetry)

The treatment that follows in this section is as first given by Landau [34] and is presented here as an aid in the development of the symmetry theory. It is now known that renormalization techniques, are required for a correct quantitative approach to critical phenomena and thus that the treatment given here should not be considered to correctly describe the quantitative behavior of G, S, and C_p in the neighborhood of the critical point. In spite of this shortcoming of the Landau theory it is generally accepted that the symmetry aspects of the theory as developed by Landau [34] are valid.

An essential feature of the theory is the consideration of the behavior of G in the region of $\eta \neq \eta^{eq}$. This consideration is accomplished through the expansion of G in a Taylor's series in η for the cases of first- or second-order transitions:

$$G = G^0 + \alpha\eta + A\eta^2 + B\eta^3 + C\eta^4 + ... \tag{VII.3}$$

where G^0, α, A, B, . . . are evaluated $\eta = 0$. It is customary, as was originally done by Landau, to terminate the expansion at the fourth-order term (although at one point in a later section it will be useful to consider the expansion to the η^6 term). It should be remembered that, as with G itself, those coefficients which need not vanish (as discussed below) are functions of thermodynamic state, e.g., of T, P, X.

First we note that the coefficient α must vanish because $\alpha = \left.\frac{\partial G}{\partial \eta}\right|_{T_t}$ and G must be at a minimum at $\eta = 0$ if $\eta = 0$ is to be stable. Thus if the thermodynamic states at which $\eta = 0$ is a stable phase are considered $\alpha(T, P, X) = 0$ and, since the transition points are included in these states, $\alpha = 0$ at the transition points. Furthermore stability of the $\eta = 0$ phase means G vs η is concave upwards at $\eta = 0$ and thus that

$$\left(\frac{\partial^2 G}{\partial \eta^2}\right)_{\eta=0} = A \geq 0 \,. \tag{VII.4}$$

Thus we have

$$G = G^0 + A\eta^2 + B\eta^3 + C\eta^4 \tag{VII.5}$$

with $A \geqq 0$. Clearly $C > 0$ or else G goes towards minus infinity and large values of η would decrease G without bound i.e., the solid would distort without limit.

We can plot $G - G^0$ vs η and obtain the plot of Fig. VII.6 for $B > 0$ (if $B < 0$ we simply reflect the plot through the vertical axis and all considerations follow). Note that a phase transition can occur when $G - G^0 = 0$, which is shown to occur when A(T, P, X) becomes sufficiently small relative to $B^2/4C$ as follows. Solving the expansion for the nonzero roots of $G - G^0 = 0$ yields

$$A\eta^2 + B\eta^3 + C\eta^4 = 0 \tag{VII.6}$$

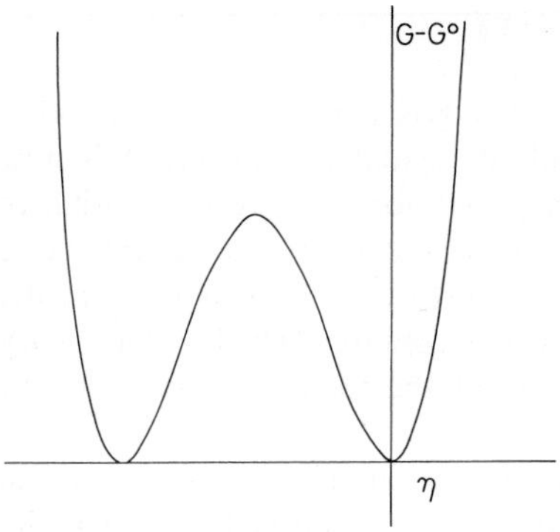

Fig. VII.7. G—G° vs η (A,B,C > 0), $B^2 = 4AC$.

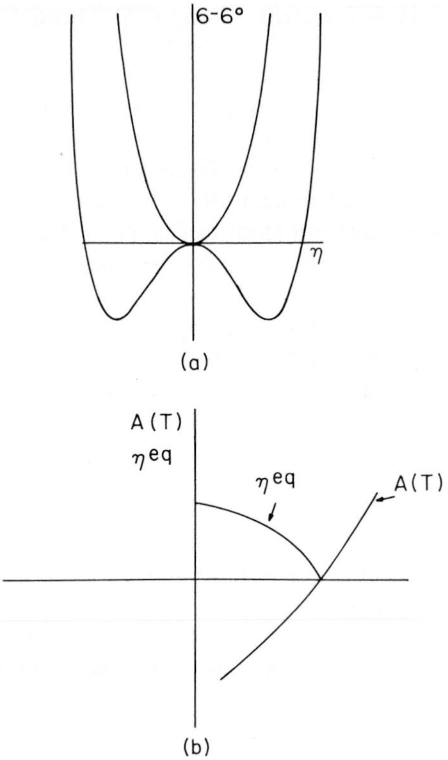

Fig. VII.8 a, b. Diagrams appropriate to $B \equiv 0$ and A(T) changes sign. **a** $G—G^0$ vs η $A > 0$ yields single minimum and $A < 0$ double minima at $\pm\eta^{eq} \neq 0$ *b* A and η^{eq} vs T showing $A > 0 = > \eta^{eq} = 0$ and $A < 0 = > \eta^{eq} \neq 0$.

or

$$A + B\eta + C\eta^2 = 0 \tag{VII.7}$$

and thus

$$\eta = -\frac{B \pm \sqrt{(B^2 - 4AC)}}{2C} \tag{VII.8}$$

which has no real roots if $4AC > B^2$ (the situation of Fig. VII.6), which has one real root if $B^2 = 4AC$ (the situation of Fig. VII.7) and otherwise has two real roots. The case of $4AC > B^2$ thus corresponds to an absolute minimum at $\eta = 0$ (stable) and a relative minimum at $\eta \neq 0$ (metastable) and the case $4AC < B^2$ reverses the stability and metastability. The case of Fig. VII.7 (i.e., $B^2 = 4AC$) is the case of $\eta^{eq} = 0$ and $\eta^{eq} = -B/2C$ and two phases coexist in equilibrium. This is the case of a first-order phase transition. Not all first-order phase transitions can be described in this way, for it is necessary for this description to be valid that the two phases be conceptually continuously related in structure. We note however, that two phases which are so related might necessarily transform via a first-order transition.

Next we inquire into the behavior of $G - G^0$ if $B \equiv 0$, i.e., if B must vanish by symmetry (a condition which will be discussed in a later section), and is therefore zero at all T, P, X points. In this case

$$G = G^0 + A\eta^2 + C\eta^4 \tag{VII.9}$$

and we have the two possibilities shown in Fig. VII.8).

In this case if $A > 0$ there is a single minimum at $\eta^{eq} = 0$ and if $A < 0$ there are two minima at $\eta^{eq} \neq 0$ (recall that $\pm\eta$ correspond to equivalent distortions). Thus if A(T, P, X) goes through zero with changing thermodynamic state η^{eq} will in this case continuously change from zero to nonzero values (see Fig. VII.8b). This is the condition for a second-order phase transition and we see from the above it can occur only if $B \equiv 0$.

The occurrence of a minimum at $\eta^{eq} \neq 0$ for the $B \equiv 0$ case means

$$\frac{\partial G}{\partial \eta} = 2A\eta_{eq} + 4C\eta_{eq}^3 = 0 \tag{VII.10}$$

and thus

$$\eta_{eq}^2 = -\frac{A}{2C}. \tag{VII.11}$$

Substitution into $G = G^0 + A\eta^2 + C\eta^4$ yields

$$G^{eq} = G^0 - \frac{A^2}{4C}. \tag{VII.12}$$

Now, if A(T, P, X) is to vanish it must be a function of state, e.g., of T, and we can expand A in a series in T and drop all terms except the first in the neighborhood of T_t, i.e., take

$$A = \alpha(T - T_t) \tag{VII.13}$$

with $\alpha > 0$ and obtain

$$G^{eq} = G^0 - \frac{\alpha^2(T - T_t)^2}{4C}. \tag{VII.14}$$

By thermodynamics $\left(\frac{\partial G}{\partial T}\right)_P = -S$ and thus

$$S^{eq} = S^0 + \frac{\alpha^2(T - T_t)}{2C} \tag{VII.15}$$

from which, as desired, $\Delta S = 0$ at $T = T_t$.

Furthermore $\left(\frac{\partial S}{\partial T}\right)_P = \frac{C_p}{T}$ and thus

$$\frac{C_p^{eq}}{T} = \frac{C_p^0}{T} + \frac{\alpha^2}{2C} \tag{VII.16}$$

and

$$\lim_{T \to T_t} \Delta C_p = \frac{\alpha^2 T_t}{2C}, \tag{VII.17}$$

and there is a discontinuity in C_p at the transition temperature.

Note that taking $A = \alpha(T - T_t)$ implies that the symmetrical ($\eta = 0$) form is the high temperature form. It is a fact that this is usually, if not always, the case for a second-order phase transition. If $A = \alpha(T_t - T)$ the other case (symmetrical form stable at $T < T_t$) would occur. In this case

$$S^{eq} = S^0 - \frac{\alpha^2(T_t - T)}{2C} \tag{VII.18}$$

and if we consider $T > T_t$ we find $S^{eq} > S^0$. This corresponds to the physically awkward situation of a larger entropy for the ordered or distorted solid than for the $\eta = 0$ cases and therefore, because of the relationship $S = -\left(\frac{\partial G}{\partial T}\right)_P$, to a G vs T diagram as shown in Fig. VII.9. It follows that the observation that Fig. VII.5 is more frequently (or always) and Fig. VII.9 less frequently (or never) correct is correlated with the physically reasonable correlation of the greater entropy with the symmetrical form.

For example, when Fig. VII.5 is correct the transition from unstable $\eta = 0$ to stable $\eta \neq 0$ at $T < T_t$ corresponds to a decrease in entropy whereas if

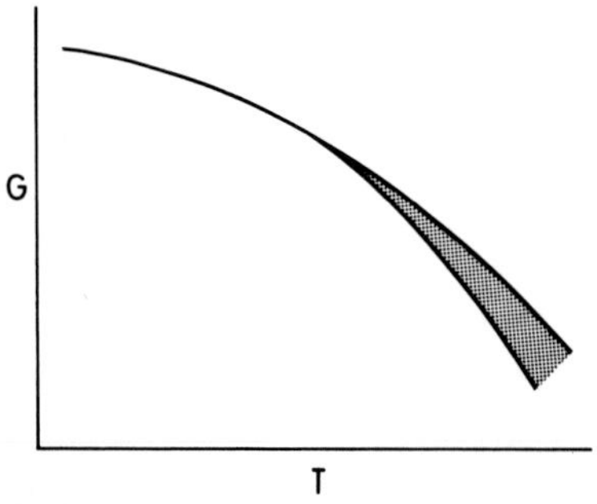

Fig. VII.9

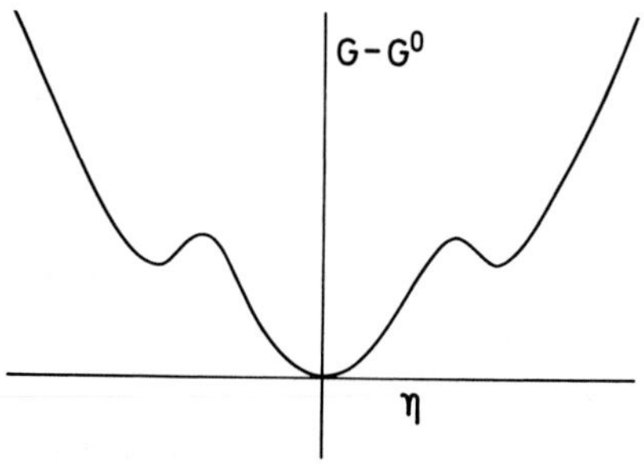

Fig. VII.10

Fig. VII.9. G vs T for $A = \alpha\,(T_t - T)$ i.e., low-symmetry form stable at high temperature.
Fig. VII.10. $G - G^\circ$ vs. η with terms to η^6. $B \equiv 0$, $C < 0$, $A, E > 0$.

Fig. VII.9 were correct the transition from unstable $\eta = 0$ to stable $\eta \neq 0$ at $T > T_t$ would occur with an increase in entropy. Such a change as implied by Fig. VII.9 would not correspond to the change in configurational entropy in CuZn or the expected change in the phonon density of states for VS. In fact, for both CuZn and VS the high temperature forms are the symmetrical ones.

VII.4 Landau Theory, With Consideration of Symmetry, Applied to the NiAs-Type To MnP-Type Phase Transition

The first condition of Landau is: in order that two crystalline structures be related by a second-order phase transition the space groups of the structures must be in the relation of a group and a subgroup. This follows from the continuity of the structure change — symmetry elements can be destroyed at a state point by infinitesimal changes of occupancy or position, but simultaneous formation of symmetry operations is not possible because the creation of symmetry elements requires definite movement of atoms which can not be accomplished by changes which are arbitrarily small in magnitude. For example the vectors of Fig. VII.1 destroy the hexagonal symmetry and maintain the vertical mirror plane to which they are parallel no matter what their magnitude. However to simultaneously create new symmetry elements in the structure would require specific, concerted movements, and could in no case be accomplished by vectors of length arbitratily close to zero.

Therefore all of the symmetry elements in the low symmetry structure are present in the high symmetry structure and since, by presumption, both the high and low symmetry structures have symmetry operations which necessarily form groups, it follows that the symmetry operations of the low symmetry structure are a subgroup of those of the high symmetry structure. We note in passing that use of the words high and low in reference to the symmetries of the structures depends upon this relationship, for only if two groups are in the relationship of a group and a subgroup can it be meaningfully said that one structure is "more symmetrical" than the other.

In the case of the phase transition from NiAs ($P6_3/mmc$) to MnP (Pcmn), the symmetry operations are in a group-subgroup relation. This can be seen in a number of ways: by consulting the tables of Neubuser and Wondratscheck [40] on groups and subgroups, by comparing figures in the International Tables [41] for Crystallography, Vol. 1 or by a detailed comparison of the symmetry operations of the two groups.

The essential symmetry elements so identified in Column 3 of Table VII.1 are present in both Pcmn and $P6_3/mmc$ and the others are lost in the transition (cf. Fig. VII.1 with the origin at the metal atom position in the hexagonal structure).

From the table it is clear that the symmetry operations of $P6_3/mmc$ fall into two categories: those that take x into itself or its negative and those that do not. Those that take x into $\pm x$ have β parts in $g_0(\mathbf{a}^*/2)$. If we examine the third column of the table we find that symmetry operations of the distorted structure are only among those that carry x into itself, or take x into $-x$ and involve

Table VII.1

Symmetry Operations of $P6_3/mmc$	Transformation of x, y, z	Present in distorted structure?	cos (2πz) sin (πx)
$\{\varepsilon \mid 000\}$	x, y, z	yes	+1
$\{C_{6z} \mid 00\frac{1}{2}\}$	$x - y, x, z + \frac{1}{2}$		
$\{C_{6z}^2 \mid 000\}$	$\bar{y}, x - y, z$		
$\{C_{2z} \mid 00\frac{1}{2}\}$	$\bar{x}, \bar{y}, z + \frac{1}{2}$	yes	+1
$\{C_{6z}^4 \mid 000\}$	$y - x, \bar{x}, z$		
$\{C_{6z}^5 \mid 00\frac{1}{2}\}$	$y, y - x, z + \frac{1}{2}$		
$\{C_{2y} \mid 000\}$	$\bar{x}, y - x, \bar{z}$		−1
$\{C_{2(x-y)} \mid 00\frac{1}{2}\}$	$\bar{y}, \bar{x}, \bar{z} + \frac{1}{2}$		
$\{C_{2x} \mid 000\}$	$x - y, \bar{y}, \bar{z}$		
$\{C_{2(2x+y)} \mid 00\frac{1}{2}\}$	$x, x - y, \bar{z} + \frac{1}{2}$		−1
$\{C_{2(x+y)} \mid 000\}$	$y, x, \bar{z}$		
$\{C_{2(x+2y)} \mid 00\frac{1}{2}\}$	$y - x, y, \bar{z} + \frac{1}{2}$		
$\{i \mid 000\}$	$\bar{x}, \bar{y}, \bar{z}$		−1
$\{C_{6z}^- \mid 00\frac{1}{2}\}$	$y - x, \bar{x}, \bar{z} + \frac{1}{2}$		
$\{C_{6z}^{-2} \mid 000\}$	$y, y - x, \bar{z}$		
$\{\sigma_z \mid 00\frac{1}{2}\}$	$x, y, \bar{z} + \frac{1}{2}$		−1
$\{C_{6z}^{-4} \mid 000\}$	$x - y, x, \bar{z}$		
$\{C_{6z}^{-5} \mid 00\frac{1}{2}\}$	$\bar{y}, x - y, \bar{z} + \frac{1}{2}$		
$\{\sigma_y \mid 000\}$	$x, x - y, z$	yes	+1
$\{\sigma_{x+y} \mid 00\frac{1}{2}\}$	$y, x, z + \frac{1}{2}$		
$\{\sigma_x \mid 000\}$	$y - x, y, z$		
$\{\sigma_{2x+y} \mid 00\frac{1}{2}\}$	$\bar{x}, y - x, z + \frac{1}{2}$	yes	+1
$\{\sigma_{x-y} \mid 000\}$	$\bar{y}, \bar{x}, z$		
$\{\sigma_{x+2y} \mid 00\frac{1}{2}\}$	$x - y, \bar{y}, z + \frac{1}{2}$		

translation by **c**/2. These are $\{\varepsilon \mid 000\}$, $\{C_2 \mid 00\frac{1}{2}\}$ $\{\sigma_{2x+y} \mid 00\frac{1}{2}\}$ and $\{\sigma \mid 000\}$). The symmetry operations that carry x into −x or involve translation by **c**/2 are not present in the distorted structure as they stand in the table, however they are present if translation by **a** is added (e.g., $\{i \mid 000\}$ is lost but $\{i \mid 100\}$ is not), and $\{C_2 \mid 10\frac{1}{2}\}$, $\{C_2 \mid 100\}$, $\{i \mid 100\}$ and $\{\sigma \mid 10\frac{1}{2}\}$ are found in the distorted structure. These symmetry operations form the space group Pcmn (minus the pure translations, which are also present in the distorted structure as is clear from Fig. VII.1), and thus the space group of the distortion is Pcmn and it is a subgroup of $P6_3/mmc$.

The second condition of Landau theory is that the distortion should correspond to a single irreducible representation of the space group of higher symmetry. This is so because, the coefficients in the expansion of G in powers of η (the A, B, C, . . .) can be associated with a single irreducible representation. That is, a coefficient A corresponding to one irreducible representation and another, A′, corresponding to another, are not equal by symmetry and therefore will be different functions of thermodynamic state and will be equal at most at isolated state points. Accordingly two A functions will not vanish at a succession of state points and behavior such as shown in Fig. VII.2a will not be observed. Thus, in general, the behavior of Fig. VII.2a corresponds to a sign change of one A, and the distortion is the one appropriate to the irreducible representation to which that A corresponds. This point will be made explicit in later sections.

An examination of the 4th column of Table VII.1 shows the behavior of the function $\varphi = \cos(2\pi z)\sin(\pi x)$ under the operations that take x into $\pm x$. It follows that φ_1 is a basis function for a small irr. rep. at **a**/2. All of the operations that remain in the distorted structure take φ_1 into φ_1, and those that are lost take φ_1 into $\pm\varphi_2$ ($= \cos(2\pi z)\sin(\pi y)$), $\pm\varphi_3$ ($= \cos(2\pi z)\sin\pi(x - y)$) or $-\varphi_1$. Accordingly the distortion corresponds to a single irreducible representation of $P6_3/mmc$, and φ_1 is one of the basis functions for this representation. Note that if x in φ_1 is replaced by $x + 1$ (translation by **a**) then the sign of the function is changed, and thus those operations which take φ_1 into $-\varphi_1$ take φ_1 into φ_1 when **a** is added to the translational part.

The significance of the correspondence of the distortion with the irreducible representation is, as will be shown, that there exists an electron density function $\Delta\varrho$, which yields $\varrho^{\text{low symmetry}}$ when added to $\varrho^{\text{high symmetry}}$ and which is a basis function for a single irreducible representation which is symmetric with respect to symmetry operations of the high symmetry group which remain and antisymmetric with respect to those that are lost. Thus $\Delta\varrho$ destroys and maintains the appropriate symmetry operations when added to ϱ^{high}. The significance of the functions φ_2 and φ_3 is that they are rotated (by 120° and 240°, respectively) φ_i's which correspond to alternative Pcmn distortions in the symmetrically equivalent directions of the hexagonal lattice.

The third condition of Landau is that no third-order combination of the basis functions be invariant under the symmetry operations of the group. As will be shown below this assures that $B \equiv 0$. It is possible to consider the operation of the symmetry operations on the set of functions φ_1, φ_2, φ_3, for example $\{C_6^5 \mid 00\tfrac{1}{2}\}$ which takes x, y, z into y, $y - x$, $z + \frac{1}{2}$ and thus φ_1 into $-\varphi_2$, φ_2 into φ_3 and φ_3 into $-\varphi_1$, i.e.

$$\begin{pmatrix} 0 & \bar{1} & 0 \\ 0 & 0 & 1 \\ \bar{1} & 0 & 0 \end{pmatrix} \begin{pmatrix} \phi_1 \\ \phi_2 \\ \phi_3 \end{pmatrix} = \begin{pmatrix} -\phi_2 \\ \phi_3 \\ -\phi_1 \end{pmatrix}. \qquad \text{(VII.19)}$$

In this fashion we can generate a set of 3×3 matrices which multiply like the symmetry operations of $P6_3/mmc$, i.e., a 3×3 representation (which is irreducible) of the space group with the basis functions φ_1, φ_2 and φ_3. Although the φ's are useful for the consideration of symmetry, they are not electron density functions, however such functions with the same symmetries could be obtained (i.e., $\Delta\varrho$'s which are also basis functions for the irreducible representation) for example by calculating ϱ for NiAs-type and for MnP-type and taking the difference with the 3 different orientations of the MnP-type structure.

A general distortion corresponding to this irr. rep. is given by

$$\varrho^{high} - \varrho^{low} = C_1(\Delta\varrho_1) + C_2(\Delta\varrho_2) + C_3(\Delta\varrho_3)\,, \qquad \text{(VII.20)}$$

and the MnP-type distortion that doubles the a axis corresponds to $C_1 \neq 0$ and $C_2 = C_3 = 0$.

Since $\varrho^{low} \to \varrho^{high}$ as the C_i's $\to 0$ it is possible to expand G for the general distortion about G^0 for the symmetrical structure:

$$G = G^0 + Af^{\langle 2\rangle}_{(C_i)} + Bf^{\langle 3\rangle}_{(C_i)} + Cf^{\langle 4\rangle}_{(C_i)} \qquad \text{(VII.21)}$$

where, as previously discussed, there is no first-order term, A is the coefficient of the second-order terms corresponding to this irr. rep. (and is the same for C_1^2, C_2^2 and C_3^2 because distortions corresponding to $\Delta\varrho_1$, $\Delta\varrho_2$ and $\Delta\varrho_3$ are equivalent by symmetry) and $f^{\langle 2\rangle}_{(C_i)}$ is the second-order combination of the C_i's that is invariant under the symmetry operations of $P6_3/mmc$, etc.

This point concerning invariance requires further amplification. Consider $\Delta\varrho = C_1(\Delta\varrho_1) + C_2(\Delta\varrho_2) + C_3(\Delta\varrho_3)$ and the effect of any of the symmetry operations of $P6_3/mmc$ upon $\Delta\varrho$. For example $\{i \mid 000\}$ takes $C_1(\Delta\varrho_1) + C_2(\Delta\varrho_2) + C_3(\Delta\varrho_3)$ into $-C_1(\Delta\varrho_1) - C_2(\Delta\varrho_2) - C_3(\Delta\varrho_3)$, i.e., we can write

$$\begin{pmatrix} \bar{1} & 0 & 0 \\ 0 & \bar{1} & 0 \\ 0 & 0 & \bar{1} \end{pmatrix} \begin{pmatrix} C_1 \\ C_2 \\ C_3 \end{pmatrix} = \begin{pmatrix} -C_1 \\ -C_2 \\ -C_3 \end{pmatrix} \qquad \text{(VII.22)}$$

and similarly we can find 3×3 matrices (multiplying the C_i column matrix) for all other operations. Thus the C_i's form a basis for the same irr. rep. as do the φ_i's. The sum $C_1^2 + C_2^2 + C_3^2$ is invariant under all symmetry operations of $P6_3/mmc$ since such operations only permute and change the signs of the C_i's. Since G for any symmetrically equivalent distortion of $P6_3/mmc$ must be the same, G is invariant under symmetry operations and so too must be the terms of G. Accordingly $C_1^2 + C_2^2 + C_3^2$ is a possible second-order term in this expansion, however $C_1C_2 + C_2C_3 + C_1C_3$ is not because under some symmetry operations of $P6_3/mmc$ this term is not invariant. We find that there is only one invariant of second-order, namely $\Sigma\, C_i^2$, and in fact such terms are generally the only second order invariants as shown by a theorem of group theory.

Attempts to find third-order invariants fail e.g., $C_1C_2C_3$ goes into $-C_1C_2C_3$ under a number of operations of $P6_3/mmc$ and also $C_1^3 + C_2^3 + C_3^3$ varies etc. Thus there is no third-order invariant ($f^{\langle 3\rangle}_{(C_i)} \equiv 0$) and no third order term. As regards fourth-order invariants examination shows that $C_1^4 + C_2^4 + C_3^4$ and $C_1^2C_2^2 + C_2^2C_3^2 + C_1^2C_3^2$ are both invariant and are independent of each other, i.e., no symmetry operation carries one into the other. Thus

$$f^{\langle 4\rangle}_{(C_i)} = C_1^4 + C_2^4 + C_3^4 + K(C_1^2C_2^2 + C_2^2C_3^2 + C_1^2C_3^2) \qquad \text{(VII.23)}$$

where $K = K(T, P, X)$.

At this point it is convenient to introduce a definition of γ_i:

$$\gamma_i^2 = C_i^2/\Sigma\, C_i^2 \qquad \text{(VII.24)}$$

(i.e., the γ_i's are normalized C_i's in the sense that $\Sigma\, \gamma_i^2 = 1$) and the definition of $\eta = \Sigma\, C_i^2$. It follows that

$$G = G^0 + A\eta^2 + C[\gamma_1^4 + \gamma_2^4 + \gamma_3^4 + K(\gamma_1^2\gamma_2^2 + \gamma_2^2\gamma_3^2 + \gamma_1^2\gamma_3^2)]\, \eta^4 \qquad \text{(VII.25)}$$

or

$$G = G^0 + A\eta^2 + C'\eta^4 \qquad \text{(VII.26)}$$

where $C' = C\,(\gamma_1^4 + \gamma_2^4 + \gamma_3^4 + K(\gamma_1^2\gamma_1^2 + \gamma_2^2\gamma_3^2 + \gamma_1^2\gamma_3^2)]$, and we return to the form of the G vs. η expression used in the preceeding section.

The effort expended above to include symmetry information in the expansion of G has resulted in additional structural information which is contained in $G(\gamma_i)$. Namely, stable structures can result only when $G(\gamma_i)$ is at a minimum. We thus seek the minima of $G(\gamma)$ subject to the restraint $\Sigma\, \gamma^2 = 1$. This amounts to minimizing $\gamma_1^4 + \gamma_2^4 + \gamma_3^4 + K(\gamma_1^2\gamma_2^2 + \gamma_2^2\gamma_3^2 + \gamma_1^2\gamma_3^2)$ subject to this restraint. This can be accomplished using Lagrange's method of undetermined multipliers. However since

$$\gamma_1^2 + \gamma_2^2 + \gamma_3^2 = 1 \qquad \text{(VII.27)}$$

it follows that

$$\gamma_1^4 + \gamma_2^4 + \gamma_3^4 + 2(\gamma_1^2\gamma_2^2 + \gamma_2^2\gamma_3^2 + \gamma_1^2\gamma_3^2) = 1 \qquad \text{(VII.28)}$$

and thus that the function to be minimized can be written

$$1 + (K - 2)\,(\gamma_1^2\gamma_2^2 + \gamma_2^2\gamma_3^2 + \gamma_1^2\gamma_3^2)\,. \qquad \text{(VII.29)}$$

This function is minimized if $\gamma_1^2\gamma_2^2 + \gamma_2^2\gamma_3^2 + \gamma_1^2\gamma_3^2$ is maximized when $K - 2 < 0$ and if $\gamma_1^2\gamma_2^2 + \gamma_2^2\gamma_3^2 + \gamma_1^2\gamma_3^2$ is minimized when $K - 2 > 0$, i.e., there are two solutions, one stable if $K < 2$, the other if $K > 2$. The smallest value that $\gamma_1^2\gamma_2^2 + \gamma_2^2\gamma_3^2 + \gamma_1^2\gamma_3^2$ can have is zero and it has this value if if $\gamma_i = 1$ and $\gamma_2 = \gamma_3 = 0$. This corresponds to the MnP-type distortion and we have now found that this is a stable solution.

A small amount of playing with numbers reveals that $\gamma_1^2\gamma_2^2 + \gamma_2^2\gamma_3^2 + \gamma_1^2\gamma_3^2$ is maximized if $\gamma_1 = \gamma_2 = \gamma_3 = 1/\sqrt{3}$ and thus this is the stable solution when $K < 2$. The structure to which this solution corresponds can be revealed by considering the vectors corresponding to the distortions of $\Delta\varrho_1$, $\Delta\varrho_2$ and $\Delta\varrho_3$ (as has been done for $\Delta\varrho_1$ in Fig. VII.1) and adding the vectors. The result is a hexagonal structure with a and b doubled relative to the NiAs-type. This structure is known for the low-temperature form of NbS. The high-temperature form of NbS is MnP-type, and thus it seems likely that, at least in principle, a yet higher form (NiAs-type) exists and that $K = 2$ somewhere in the homogeneity range of NbS. In summary, we have shown that the Landau theory provides relationships between structures (e.g., between NbS (l.t.) and MnP-type) which correspond to possible stable solutions for the same irr. rep. of a space group, and indicates the existence of structure types that might be sought experimentally.

In closing this section it is important to remark that even if $B \equiv 0$ a transition still need not be second-order. In order to consider this possibility it is necessary to consider terms to 6th order (5th and higher odd order invariants do not exist):

$$G = G^0 + A\eta^2 + C\eta^4 + E\eta^6 \, . \qquad \text{(VII.30)}$$

Now if $A, E > 0$ and $C < 0$ we have $G - G^0$ vs. η as shown in Fig. VII.10. The minima at $\eta \neq 0$ are reminiscent of that $B \neq 0$ and the behavior is similar, i.e., a first-order transition necessarily results.

VII.5 General Development of the Landau Theory With Consideration of Symmetry

The electron density ϱ, of the low symmetry phase is expressed as a linear combination of basis functions for the irr. rep. of the space group of higher symmetry.

$$\varrho = \sum_\alpha \sum_i C_i^\alpha (\Delta\varrho_i)^\alpha \qquad \text{(VII.31)}$$

where α labels the irr. rep. and i the degenerate basis functions for a given irr. rep. That this may be done is a theorem of group theory, and the procedure whereby the coefficients would be found once the $(\Delta\varrho_i)^\alpha$'s were known is called projection.

Among the irr. reps. there is always one corresponding to ϱ^0 called the totally symmetric which has the symmetry of the high-symmetry space group. This term is removed from the sum

$$\varrho = \varrho^0 + \sum_\alpha{}' \sum_i C_i^\alpha (\Delta\varrho_i)^\alpha \, . \qquad \text{(VII.32)}$$

Since $\varrho \to \varrho^0$ as C_i's $\to 0$, G, the Gibbs free energy of a general distortion is expanded in the C_i's

$$G = G^0 + A_1\eta_1^2 + A_2\eta_2^2 + \ldots \qquad \text{(VII.33)}$$

where we make use of the fact that the only second-order invariant corresponding to each irr. rep. is $\Sigma\, C_i^2 = \eta_\alpha$. Now continuous distortion occurs when one of the

A_α's changes sign and, since the A_α's are independent, with very high probability only one A_α will change sign at a given state point. Even if a point at which two A_α's vanished did occur it would not be the case that two A_α's would change sign at a succession of state points such as represented by the line of Fig. VII.2a. Accordingly a second-order phase transition that occurs at a succession of state points (T vs. X or T vs. P, etc.) will correspond to a single irr. rep. of the space group of higher symmetry.

Thus we consider only the terms for a single irr. rep. i.e., defining $\gamma_i^2 = C_i^2/\eta$,

$$G = G^0 + A\eta^2 + Bf^{\langle 3\rangle}(\gamma_i)\,\eta^3 + Cf^{\langle 4\rangle}(\gamma_i)\,\eta^4 \qquad \text{(VII.34)}$$

to terms of fourth order, where $f^{\langle n\rangle}(\gamma_i)$ is an nth order homogeneous function of the γ_i's. We have seen that in order for the transition to be second-order the coefficient of η^3 must vanish. Since symmetry operations carry

$$\varrho = \varrho^0 + \sum_i C_i(\Delta\varrho_i) \qquad \text{(VII.35)}$$

into equivalent electron densities it follows that G must be invariant under symmetry operations (equivalent distortions have the same G). Such symmetry operations alter the coefficients of the basis functions, and thus one must examine whether $f^{\langle n\rangle}(\gamma_i)$ exists and what its form is for various n's. For $n = 1$ there exists no invariant except for the totally symmetric irr. rep. which would correspond to no change in symmetry. For $n = 2$, $f^{\langle n\rangle}(\gamma_i) = \Sigma\,\gamma_i^2$ is always the only invariant that need be considered. For $n = 3$ there may or may not be invariants, and this must be investigated. It is necessary for a second-order phase transition to occur that there not be third-order invariants. For $n = 4$, $(\Sigma\,\gamma_i^2)^2$ is always one invariant, however the number of independent 4th order invariants must be determined. If there is more than one then there is more than one possible distortion corresponding to the irr. rep. To determine the stable solutions it is necessary to minimize $G(\gamma)$ subject to the restraint $\Sigma\,\gamma^2 = 1$.

VII.6 Landau's 4th Condition [42]

We have seen that A, the lead coefficient in $G - G^0$, corresponds to a given irr. rep. and that the irr. reps. correspond to particular **k** vectors. It follows that A depends upon **k** and in fact can be expanded in $\delta\mathbf{k}$ about a given **k**, i.e.,

$$A(\mathbf{k} + \delta\mathbf{k}) = A(\mathbf{k}) + \boldsymbol{\alpha}\cdot\delta\mathbf{k} + \ldots \qquad \text{(VII.36)}$$

where $\boldsymbol{\alpha}$ must be a vector and the first order term is a dot product because A is a scalar. In general $\boldsymbol{\alpha}(T, P, X)$ will exist and be nonzero and therefore some $\delta\mathbf{k}$ for which $\boldsymbol{\alpha}\cdot\boldsymbol{\alpha}\mathbf{k} < 0$ will necessarily exist, and $A(\mathbf{k} + \delta\mathbf{k}) < A(\mathbf{k})$ is assured. However if this is the case then a distortion at point k is not allowed since there exists a more stable point in the neighborhood of **k** (namely at $\mathbf{k} + \delta\mathbf{k}$) and thus the whole consideration given in the earlier sections to the vanishing of A is irrelevant since the distortion is to a different k point with a different $g_0(k)$ and a different set of irr. reps. If nondegenerate small representations are considered the treatment given above can be rescued if $\boldsymbol{\alpha} \equiv 0$, i.e., if there is no vector invariant for symmetry reasons. That is, suppose that the symmetry of reciprocal space is such that $\delta\mathbf{k}$ and $\beta\delta\mathbf{k}$ have the same A (i.e., suppose β is in $g_0(\mathbf{k})$). Then

$$\boldsymbol{\alpha} \cdot \delta\mathbf{k} = \boldsymbol{\alpha} \cdot \beta\delta\mathbf{k} \tag{VII.37}$$

or

$$\boldsymbol{\alpha} \cdot \delta\mathbf{k} = \beta^{-1}\boldsymbol{\alpha} \cdot \delta\mathbf{k} \tag{VII.38}$$

for all β and $\delta\mathbf{k}$, i.e., $\boldsymbol{\alpha} = \beta^{-1}\boldsymbol{\alpha}$. This will be possible with $\boldsymbol{\alpha} \neq 0$ for some special cases, e.g., a single axis ($\boldsymbol{\alpha}$ lies along axis) or plane ($\boldsymbol{\alpha}$ lies in plane). However if β^{-1} is inversion, or if there are two β's such as an intersecting plane and axis or intersecting pair of axes, then the above can be true only if $\alpha \equiv 0$ and a second-order phase transition to $\mathbf{k}$ is possible. The fourth condition of Landau then requires that there be an inversion or a pair of intersecting axes or an intersecting axis and mirror in $g_0(k)$, etc. in order that a second-order phase transition be possible. Note that the case of $P6_3/mmc$ at $\mathbf{k} = \mathbf{a}^*/2$ meets this condition since i is included in D_{2h}.

However, an implicit assumption in the above discussion is that the unsymmetrical structure has space group symmetry, and thus that the transition is to a $\mathbf{k}$ point which corresponds to a given fixed periodicity (such as $k = \mathbf{a}^*/2$ which doubles the periodicity in the $\mathbf{a}$ direction). Suppose instead [42] that the transition is to a variable $\mathbf{k}$ (e.g., to $\mathbf{k} + \delta\mathbf{k}$ with $\delta\mathbf{k}$ variable). Then the transition can occur to a point at which $\boldsymbol{\alpha}(T, P, X)$ happens to be zero, and that point will change, for example with changing somposition or pressure. Such a case is an incommensurate phase (since $\delta\mathbf{k}$ is not an integral submultiple of a reciprocal lattice vector) and the characteristic length and direction of the incommensurate portion of the structure varies with state. In this case the 4th condition, which is appropriate only for transitions to a given $\mathbf{k}$ point, is not applicable.

VII.7 Problems

1. a. Show that the symmetry operations of $P6_3/mmc$ with β parts in $g_0(\mathbf{a}^*/2)$ take x into $\pm$x; b. Show that those that take x into $-$x take x into $+$x when combined with translation by $\mathbf{c}/2$.
2. Find the 3×3 matrices that represent $C_{6z} \mid 00\tfrac{1}{2}$, $C_{2z} \mid 00\tfrac{1}{2}$, $C_{2x} \mid 000$ and $i \mid 000$ in the 3 dimensional irr. rep. of NiAs symmetry to which the MnP distortion corresponds.
3. Find a symmetry operation of $P6_3/mmc$ to which $C_1C_2 + C_2C_3 + C_1C_3$ is not invariant.
4. Find a symmetry operation of $P6_3/mmc$ to which $C_1^3 + C_2^3 + C_3^3$ is not invariant.
5. Create a drawing of the $\gamma_1 = \gamma_2 = \gamma_3 = 1/\sqrt{3}$ solution for the distortion corresponding to the irr. rep. of NiAs symmetry to which MnP correspond.

6. Make a plot of a function $G = G^0 + A\eta^2 + C\eta^4 + E\eta^6$ with specified values of A, C and E which exhibits a first order transition.
7. Show that no vector invariant exists if a four-fold improper axis is in the point group of the wave vector.
8. Consider the phase transition from ordered Cu_3Au to disordered fcc type from the point of view of Landau theory and decide whether the transition can occur as a second-order phase transition.
9. Show that a first-order phase transition between two crystalline modifications with the same chemical composition in a c = 2 system in which the solids exhibit nonstoichiometry (which, in principle, is true of all solids, as discussed in Chap. IV) occurs at a maximum or minimum in T vs X.
10. Consider a transition in which a bcc structure destorts to form a bct (body-centered tetragonal) structure by the increase (or decrease) in $|\mathbf{c}|$ relative to $|\mathbf{a}| = |\mathbf{b}|$. Apply Landau's theory and determine whether the transition can occur as a second-order phase transition.
11. Find the space group that would result if a distortion of $P6_3/mmc$ at $\mathbf{a}^*/2$ corresponded to the small representation:

ε	C_{2z}	C_{2y}	$C_{2(2x+y)}$	i	6_z	6_y	$6_{2(2x-y)}$
1	1	1	1	−1	−1	−1	−1

12. Find the **k** vector(s) to which the ordering of vacancies in the metal-atom positions of an NaCl structure with a = 5.17 Å yields a trigonal structure of vacancies on a sublattice with a = 3.655, c = 8.95 Å. What is the symmetry of the resultant structure? What is the stoichiometry if only the sublattice is completely empty?
13. Consider a hypothetical order-disorder transition in which the atoms of the WC-type structure exchange positions to achieve a random distribution. Is this transition possible as a second-order phase transition?

Chapter VIII

Diffraction by Crystalline Solids

VIII.1 Introduction

Complex exponentials exp $i\alpha$ are ideal mathematical expressions for treating the interactions of waves. Multiplying a real vector of magnitude f by the complex exponential, exp $i\alpha$, rotates the vector by the angle α in the complex plane, i.e., (see Fig. VIII.1) $f \cdot \exp i\alpha = f \cdot \cos\alpha + i \cdot f \cdot \sin\alpha$. Therefore, a term like $f \exp i\alpha$ contains both a phase angle (α) and a magnitude (f), and thus contains

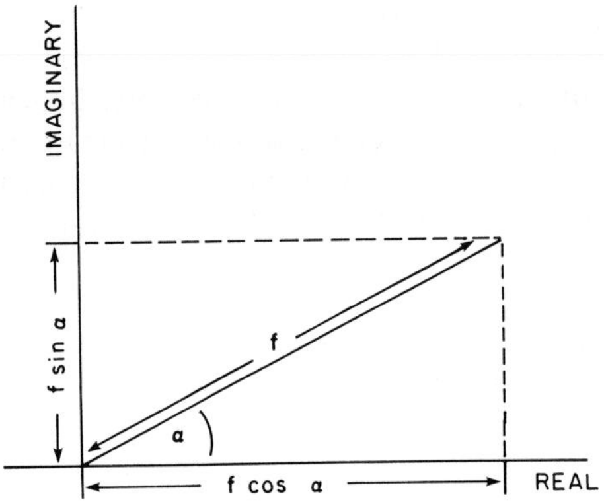

Fig. VIII.1. exp i_α as an operator which rotates a vector of magnitude f by the angle α in the complex plane.

the information required for treating wave interactions. For example, two waves differing in magnitude and phase angle interact according to (see Fig. VIII.2)

$$f \exp i\alpha = f_1 \exp i\alpha_1 + f_2 \exp i\alpha_2, \qquad \text{(VIII.1)}$$

where f is the magnitude and α the phase angle of the resultant.
and

$$|f|^2 = ff^* \qquad \text{(VIII.2)}$$

$$= (f_1 \exp i\alpha_1 + f_2 \exp i\alpha_2)(f_1 \exp -i\alpha_1 + f_2 \exp -i\alpha_2) \qquad \text{(VIII.3)}$$

$$= f_1^2 + f_2^2 + f_1 f_2 \exp i(\alpha_1 - \alpha_2) + f_2 f_1 \exp i(\alpha_2 - \alpha_1) \qquad \text{(VIII.4)}$$

$$= f_1^2 + f_2^2 + f_1 f_2 (\exp i(\alpha_1 - \alpha_2) + \exp -i(\alpha_1 - \alpha_2)) \qquad \text{(VIII.5)}$$

$$= f_1^2 + f_2^2 + 2 f_1 f_2 \cos(\alpha_1 - \alpha_2)\,. \qquad \text{(VIII.6)}$$

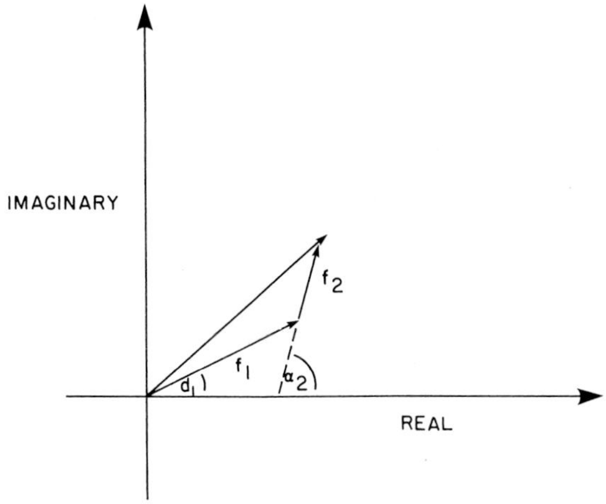

Fig. VIII.2. $f \exp i_{\alpha} = f_1 \exp i_{\alpha_1} + f_2 \exp i_{\alpha_2}$.

In the case of diffraction of radiation by crystalline matter it is assumed that each atom among a very large number (N) scatters radiation of a single ray with a given wave length, λ. Within a unit cell each atom labeled by the subscript j is assumed to have an effective scattering factor of f_j, where for X-ray and electron diffraction this factor has been corrected for interference effects within the electron density distributed on the atom.

VIII.2 Diffraction

When an origin is arbitrarily specified at a given atomic position, all other atomic positions are specified by vectors $\mathbf{r}_j + \mathbf{T}_l$ where j labels the atom position within the cell and $\mathbf{T}_l$ is a translational symmetry operation for the crystal structure. The difference in path length for radiation scattered in a given direction from the atom at the origin and that at $\mathbf{r}_j + \mathbf{T}_l$ is (see Fig. VIII.3) $\Delta L = (\mathbf{r}_j + \mathbf{T}_l) \cdot (\mathbf{k}_{out} - \mathbf{k}_{in})$ where $\mathbf{k}_{in}$ and $\mathbf{k}_{out}$ are vectors of unit length in the direction of the incoming ray (directed toward the crystal) and the detector (directed away from the crystal), respectively. Division by λ yields the difference in phase ($\Delta\varphi$) of the scattered radiation from the two atomic positions: $\Delta\varphi = (\mathbf{r}_j + \mathbf{T}_l) \cdot (\mathbf{k}_{out} - \mathbf{k}_{in})/\lambda$.

Summing over all atomic sites in a unit cell (over j) and over all N (a very large number) sites with which the ray interacts equivalent to these by translation, yields an F factor

$$F_j = \sum_l{}' \sum{}' f_j \exp\left[-\frac{2\pi i(\mathbf{r}_j + \mathbf{T}_l) \cdot (\mathbf{k}_{out} - \mathbf{k}_{in})}{\lambda}\right], \qquad \text{(VIII.7)}$$

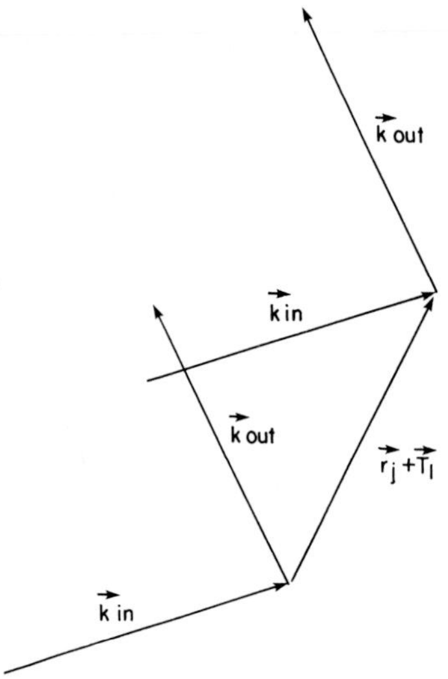

Fig. VIII.3

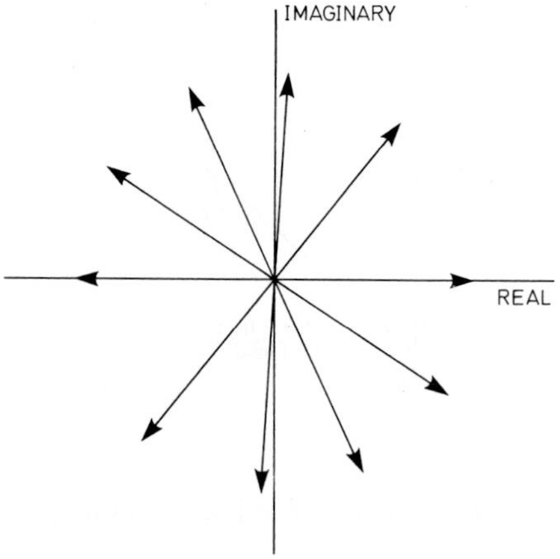

Fig. VIII.4

Fig. VIII.3. Scattering diagram for scatterers, separated by $\mathbf{r}_j + \mathbf{T}_l$.

Fig. VIII.4. Illustration that $\sum_{N=1}^{10} \exp 0.9\pi i = 0$.

the square of the magnitude ($\sqrt{FF^*}$) of which is the scattered intensity in the direction of $\mathbf{k}_{out}$. Factoring

$$F = \sum_j f_j \exp\left(-\frac{2\pi i \mathbf{r}_j \cdot (\mathbf{k}_{out} - \mathbf{k}_{in})}{\lambda}\right) \sum_{l=1}^{N}{}' \exp - \left(\frac{2\pi i \mathbf{T}_l \cdot (\mathbf{k}_{out} - \mathbf{k}_{in})}{\lambda}\right) \tag{VIII.8}$$

First we examine the lattice term:

$$\sum_{l=1}^{N}{}' \exp\left(-\frac{2\pi i \mathbf{T}_l \cdot (\mathbf{k}_{out} - \mathbf{k}_{in})}{\lambda}\right) \tag{VIII.9}$$

and consider, for example, a case, $\mathbf{T}_n$, for which

$$\frac{\mathbf{T}_n \cdot (\mathbf{k}_{out} - \mathbf{k}_{in})}{\lambda} = 0.9\,. \tag{VIII.10}$$

In this case the translations $\mathbf{T}_n$, $2\mathbf{T}_n$, $3\mathbf{T}_n$, $4\mathbf{T}_n$, ... in the sum yield phase angles 1.8π, 3.6π, 5.4π, 7.2π, which are equivalent to 1.8π, 1.6π, 1.4π, 1.2π, ...

Figure VIII.4 shows the sum $F = \sum_{l=1}^{N} \exp\left(-\frac{2\pi i \mathbf{T}_n(\mathbf{k}_{out} - \mathbf{k}_{in})}{\lambda}\right) = \exp(-1.8\pi i)$ $+\exp(-1.6\pi i) + \exp(-1.4\pi i) + \exp(-1.2\pi i) + \ldots$ for this case, and it is clear that this vector sum is zero (for each vector there is an inverse). In fact the sum will continue to be zero for very large values of N until each phase angle is very nearly an even integral of π so that each vector in the sum is almost lined up along the real axis (constructive interference). Therefore, in order that any appreciable scattering can occur it is necessary that

$$\frac{\mathbf{T}_n \cdot (\mathbf{k}_{out} - \mathbf{k}_{in})}{\lambda} = \text{an integer} \qquad \text{(VIII.11)}$$

to within a very high degree of precision. We know from Chap. V what sorts of vectors yield integral values when the dot product is taken with a translational symmetry operation, namely reciprocal lattice vectors, $\mathbf{k} = m^*\mathbf{a}^* + n^*\mathbf{b}^* + p^*\mathbf{c}^*$ where

$$\mathbf{a}^* = \frac{\mathbf{b} \times \mathbf{c}}{\mathbf{a} \cdot \mathbf{b} \times \mathbf{c}}, \qquad \text{(VIII.12)}$$

and etc. Hence the condition for diffraction will be met whenever

$$\frac{\mathbf{k}_{out} - \mathbf{k}_n}{\lambda} = \mathbf{K}, \qquad \text{(VIII.13}$$

for when this condition is met each term in

$$\sum_{l=1}^{N} \exp\left(-2\pi i \mathbf{T}_l \cdot \frac{(\mathbf{k}_{out} - \mathbf{k}_{in})}{\lambda}\right) \qquad \text{(VIII.14)}$$

is an integer, and the condition for constructive interference is met.

VIII.3 Miller Indices

A particular family of planes of points of a Bravais lattice is labeled by its crystallographic indices, hkl, where these indices can be determined as follows (refer to Fig. VIII.5):

1. Three vectors originating from a common lattice point, n**a**, m**b** and p**c** terminate at lattice points in the plane (pick the closest point to the plane for which this is true, i.e., n, m and p have no common divisor),
2. Take the inverses of the resultant intercept indices (3, 6, 5 are the intercept indices in this case, and the inverses are $\frac{1}{3}, \frac{1}{6}, \frac{1}{5}$),
3. Clear fractions by multiplying by the smallest nonzero number that yields a triple of integers (which is 30 for $\frac{1}{3}, \frac{1}{6}, \frac{1}{5}$, yielding 10, 5, 6).

If L_1 is the least common integral multiple of m and n, L_2 that of n and p and L_3 that of m and p, and $L = L_1L_2L_3$, then the multiplier of the inverse indices is (mnp)/L, i.e., the crystallographic indices are:

$$h = np/L, \; k = mp/L \text{ and } l = mn/L .$$

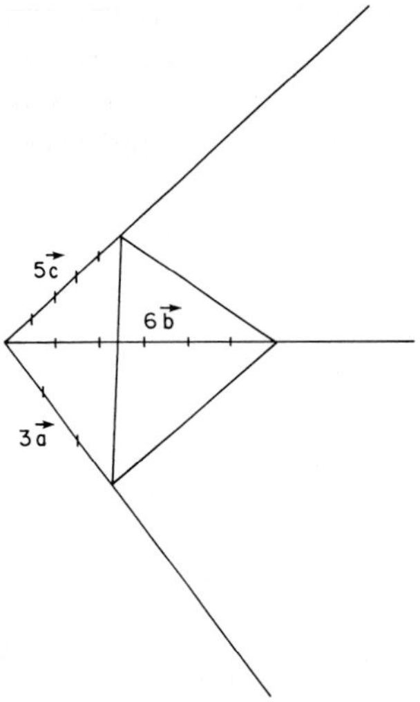

Fig. VIII.5. The 3,6,5 rational intercept plane.

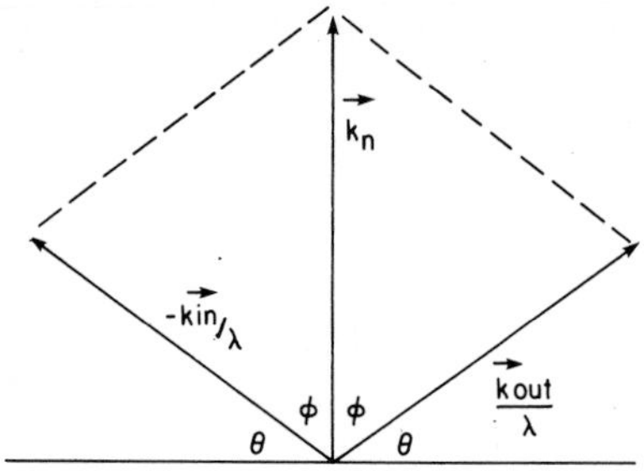

Fig. VIII.6. $\mathbf{K}_n = \dfrac{\mathbf{k}_{out} - \mathbf{k}_{in}}{\lambda}$.

Thus if m = 10, n = 6, p = 15 then the inverse intercept indices are $\frac{1}{10}$, $\frac{1}{6}$, $\frac{1}{15}$, L = 2 · 3 · 5 = 30 and h = 3, k = 5 and l = 2.

The crystallographic indices as defined here are all mutually prime and they refer only to families of planes which contain the lattice points. It is usual to generalize these indices to include, for example, as well as the 1, 1, 1 planes, the 2, 2, 2; 3, 3, 3; etc. The resultant Miller indices label those families of planes which are parallel to the h, k, l planes but with half the interplanar spacing for 2h, 2k, 2l, and with $\frac{1}{n}$th the interplanar spacing for nh, nk, nl.

To find the relationship between the reciprocal lattice and the planes of lattice points consider two vectors in a given plane of a family labeled by intercept indices m, n, p (no common multiple), namely m**a**—n**b** and m**a**-p**c**, and in order to find a vector perpendicular to the plane take the cross product between them:

$$(m\mathbf{a} - n\mathbf{b}) \times (m\mathbf{a} - p\mathbf{c}) = -mp\mathbf{a} \times \mathbf{c} - nm\mathbf{b} \times \mathbf{a} + np\mathbf{b} \times \mathbf{c} \qquad \text{(VIII.15)}$$

$$= V_{cell}(mp\mathbf{b}^* + nm\mathbf{c}^* + np\mathbf{a}^*)\,, \qquad \text{(VIII.16)}$$

$$= V_{cell} \cdot L(h\mathbf{a}^* + k\mathbf{b}^* + l\mathbf{c}^*)\,. \qquad \text{(VIII.17)}$$

Thus the reciprocal lattice vector $h\mathbf{a}^* + k\mathbf{b}^* + l\mathbf{c}^*$ is perpendicular to the planes in the family labeled by the Miller indices h, k, l.

By Eq. VIII.13 diffraction occurs when $(\mathbf{k}_{out} - \mathbf{k}_{in})/\lambda$ is a reciprocal lattice vector, and since $|\mathbf{k}_{ut}| = |\mathbf{k}_{in}|$ this reciprocal lattice vector bisects the angle between the incoming and diffracted rays. Hence (Fig. VIII.6) when this reciprocal lattice vector is $\mathbf{K}_n = h\mathbf{a}^* + k\mathbf{b}^* + l\mathbf{c}^*$ the angle of incidence with the hkl planes equals the diffraction angle, Θ. To find this angle, take the dot product of each side of

$$h\mathbf{a}^* + k\mathbf{b}^* + l\mathbf{c}^* = \frac{\mathbf{k}_{out} - \mathbf{k}_{in}}{\lambda} \qquad \text{(VIII.18)}$$

with $\mathbf{K}_n$,

$$\mathbf{K}_n \cdot \mathbf{K}_n = |\mathbf{K}_n|^2 = \frac{\mathbf{K}_n \cdot \mathbf{k}_{out} - \mathbf{K}_n \cdot \mathbf{k}_{in}}{\lambda} \qquad \text{(VIII.19)}$$

to obtain (Fig. VIII.6)

$$|\mathbf{K}_n|^2 = \frac{|\mathbf{K}_n|\, 2\cos\varphi}{\lambda} \qquad \text{(VIII.20)}$$

or

$$|\mathbf{K}_n| = \frac{2\sin\theta}{\lambda} \qquad \text{(VIII.21)}$$

Next we show that $|\mathbf{K}_n| = 1/d_{hkl}$. Consider the hkl planes (see Fig. VIII.7). The planes cut the axes in units of $\mathbf{a}/h$, $\mathbf{b}/k$ and $\mathbf{c}/l$ (i.e., there are mh = nk = pl planes between the origin lattice point and the lattice intercept plane). Furthermore, the vector $\mathbf{K}_n/|\mathbf{K}_n|$ is a unit vector perpendicular to the planes, thus

$$d = \frac{\mathbf{K}_n}{|\mathbf{K}_n|} \cdot \frac{\mathbf{a}}{h} = \frac{\mathbf{K}_n}{|\mathbf{K}_n|} \cdot \frac{\mathbf{b}}{k} = \frac{\mathbf{K}_n}{|\mathbf{K}_n|} \cdot \frac{\mathbf{c}}{l} = \frac{1}{|\mathbf{K}_n|} \,. \qquad \text{(VIII.22)}$$

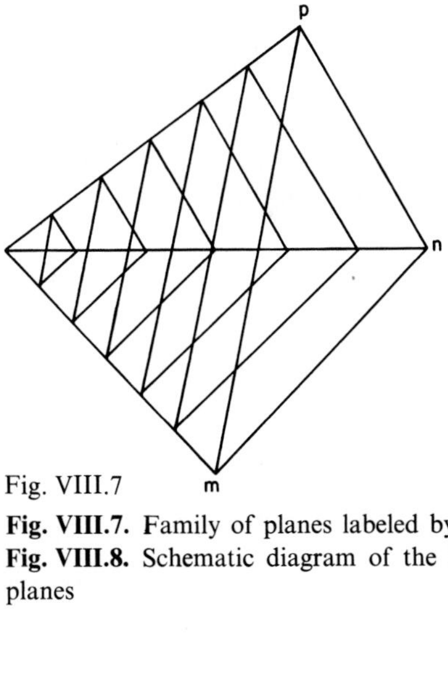

Fig. VIII.7

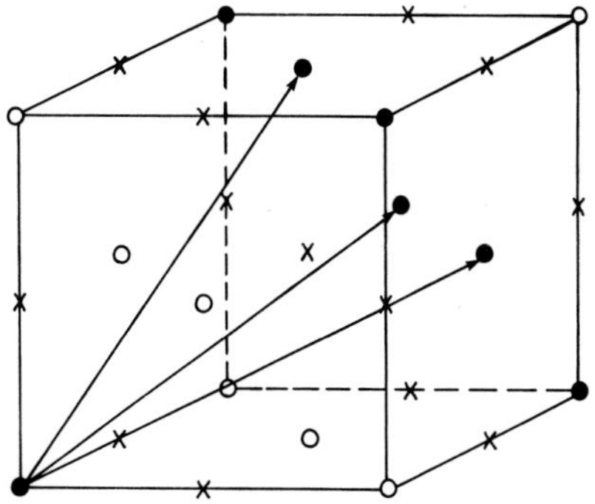

Fig. VIII.8

Fig. VIII.7. Family of planes labeled by the rational intercept indices m, n, p.

Fig. VIII.8. Schematic diagram of the $Sc_{1-x}S$ structure with vacancies in alternate 1, 1, 1 planes

Substituting into Eq. 21 yields

$$\lambda = 2d \sin\theta\,, \qquad \text{(VIII.23)}$$

the Bragg relation.

VIII.4 Structure Factor

What has been demonstrated up to this point is that the Bragg law (Eq. 23) is valid for diffraction from crystals with lattice spacings d and Bragg angle Θ regardless of how the atoms are distributed relative to the lattice points. The terms obtained from the consideration of the lattice sum are the same for each reciprocal lattice point, and the same would be true of the intensities, FF*, (corrected for Lorentz and polarization effects, for thermal motion and for absorption, extinction, etc. with which we shall not deal here) were it not for the remaining term. We now take this remaining term to be the whole of F, viz.

$$F = \sum_j f_j \exp - 2\pi i \mathbf{r}_j \cdot (h\mathbf{a}^* + k\mathbf{b}^* + l\mathbf{c}^*) \qquad \text{(VIII.24)}$$

or

$$F = \sum_j f_j \exp - 2\pi i(hx_j + ky_j + lz_j)\,, \qquad \text{(VIII.25)}$$

since it contains all of the structural scattering intensity information.

This sum is over all atoms in the unit cell, and it is this term which relates diffraction intensities to the distribution of scattering matter within the unit cell. The diffraction directions are determined by the lattice, the diffraction intensities by the distribution of scattering matter relative to the lattice points.

VIII.5 Extinctions

Consider a diffraction experiment in the case of a unit cell that is face-centered, i.e., $\frac{1}{2}$, 0, $\frac{1}{2}$; $\frac{1}{2}$, $\frac{1}{2}$, 0 and 0, $\frac{1}{2}$, $\frac{1}{2}$ are translational symmetry operations. Then for each atom labeled by j at x_j, y_j, z_j there are three others at $x_j + \frac{1}{2}$, y_j, $z_j + \frac{1}{2}$; $x_j + \frac{1}{2}$, $y_j + \frac{1}{2}$, z_j and x_j, $y_j + \frac{1}{2}$, $z_j + \frac{1}{2}$. Thus in the sum for F for each term $f_j \exp -2\pi i\,(hx_j + ky_j + lz_j)$ there are three others, i.e., the sum is made up of terms like

$$\begin{aligned} &f_j(\exp -2\pi i(hx_j + hy_j + lz_j) + \exp -2\pi i[h(x_j + \tfrac{1}{2}) + ky_j \\ &+ l(z_j + \tfrac{1}{2})] + \exp -2\pi i[h(x_j + \tfrac{1}{2}) + k(y_j + \tfrac{1}{2}) + lz_j] + \exp -2\pi i \\ &\times [hx_j + k(y_j + \tfrac{1}{2}) + l(z_j + \tfrac{1}{2})]) \qquad \text{(VIII.26)} \\ &= f_j(1 + \exp -\pi i(h + l) + \exp -\pi i(h + k) + \exp -\pi i(k + l)) \\ &\quad \times \exp -2\pi i(hx_j + ky_j + lz_j)\,. \qquad \text{(VIII.27)} \end{aligned}$$

Now if h, k, l are all even or all odd then

$$[1 + \exp -\pi i(h + l) + \exp -\pi i(h + k) + \exp -\pi i(k + l)] = 4 \tag{VIII.28}$$

and $F = 4 \Sigma f_j \exp 2\pi i(hx_j + ky_j + lz_j)$, where the sum is over the primitive unit cell, rather than the face centered one. If, on the other hand, any one of the indices h, k or l is odd while the others are even (or vice versa), then

$$[1 + \exp -\pi i(h + l) + \exp -\pi i(h + k) + \exp -\pi i(h + l)] = 0, \tag{VIII.29}$$

since among the sums h + l, h + k and k + l two will be odd and one even, leading to 2 — 2 = 0. Hence, only diffraction corresponding to lattice planes with h, k, l all even or all odd are observed from face centered cells.

An alternative view of this "extinction" effect can be obtained from the perspective of reciprocal space. A family of lattice planes corresponds to a reciprocal lattice vector and thus to a reciprocal lattice point. The description of a lattice as face-centered (i.e., with a face-centered unit cell) as opposed to the description in terms of a primitive lattice, has suggested the inclusion of lattice points on the reciprocal lattice which in fact should not have been included (and, similarly the occurence of diffraction maxima which do not occur). The extinction of the diffraction maxima corresponds to the nonpresence of the reciprocal lattice points, i.e., the correct reciprocal lattice for a face-centered lattice includes only all even or all odd integral combinations of **a***, **b*** and **c***.

The reflection conditions appropriate to end-centered, body-centered and face-centered cells are summarized in Table VIII.1.

Table VIII.1

A centered	k + l = 2n
B centered	l + h = 2n
C centered	h + k = 2n
I centered	h + k + l = 2n
F centered	h + k = 2n and k + l = 2n

Other systematic extinctions occur as the result of nonsymmorphic symmetry operations. As an example, consider the operation of $C_{2z} \mid 00\frac{1}{2}$. For each x_j, y_j, z_j there occurs $\bar{x}_j, \bar{y}_j, z_j + \frac{1}{2}$ and thus

$$F = \Sigma f_j[\exp -2\pi i(hx_j + ky_j + lz_j) + \exp -2\pi i(-hx_j - ky_j + l \times (z_j + \tfrac{1}{2})] \tag{VIII.30}$$

and for h = k = 0(001) reflections

$$F = \Sigma f_j \exp -2\pi ilz_j[1 + \exp \pi il] \tag{VIII.31}$$

which vanishes for l odd but not for l even. Hence, the condition for diffraction is 00l, l = 2n. In a similar fashion all of the various extinction conditions that lead to the knowledge of the occurrence of nonsymmorphic symmetry elements among the essential symmetry elements can be determined.

Regarding nonessential symmetry elements, note that, for example, body-centering together with a C_{2z} through the origin implies the existence of a screw axis (the operation $C_{2z} \mid \frac{1}{2}\frac{1}{2}\frac{1}{2}$) located at $x = y = \frac{1}{4}$. However, the body-centering condition $(h + k + l = 2n)$ includes the screw axis condition (00l, $l = 2n$) as a special case.

The extinction conditions present and not present can thus frequently but not always (compare I222 and $I2_12_12_1$) be used to determine the existence of essential nonsymmorphic symmetry elements.

VIII.6 Order-Disorder and Superstructure

For $Sc_{1-x}S$ at high temperatures the $(1 - x)$ N scandium atoms are distributed at random over N sites of an NaCl-type structure with the net result that X-rays, which average over a large number of sites, effectively endow each Sc site with a scattering factor $(1 - x) f_{Sc}$. The scandium atoms are located at 0, 0, 0 and the sites related to the origin by face-centering, hence the scandium contribution to F for hkl all even or all odd reflections is $4(1 - x) f_{Sc}$ (since $\exp -2\pi i(hx + ky + lz) = 1$, see preceeding section on extinction in face centered structures). The sulfur atoms are located at 0, 0, $\frac{1}{2}$ and the positions related by face centering, and thus the sulfur contribution to F for observed reflections is $4f_S \exp -\pi il$, or

$$F = 4[(1 - x) f_{Sc} + f_S \exp \pi il] . \qquad \text{(VIII.32)}$$

All the indices h, k, l are odd or even together hence l is even when h, k and l are even and

$$F = 4[(1 - x) f_{Sc} + f_S] \qquad \text{(VIII.33)}$$

and l is odd when, h, k and l are odd and

$$F = 4[(1 - x) f_{Sc} - f_S] . \qquad \text{(VIII.34)}$$

Thus, except for the various physical factors that effect intensity (Lorentz, polarization, absorption, etc.) there are two observed diffraction intensities, strong when the S and Sc sublattices are exactly in phase and weak when they are exactly out of phase, and the former occurs for h, k, l even reflections (2, 0, 0; 2, 2, 0; 2, 2, 2; etc.) and the latter for h, k, l odd (1, 1, 1; 3, 1, 1; 3, 3, 1; etc.). The effect of the vacancies is to weaken the already weak reflections since $f_{Sc} < f_S$ (roughly f_{Sc}/f_S is equal to the ratio of the number of electrons on Sc and S, i.e., about $\frac{21}{16} = 1.3$).

At lower temperatures the vacancies on the Sc sites tend to accumulate in alternate Sc containing planes along the 1, 1, 1 direction (see Fig. VIII.8). The accompanying diffraction effects can be determined by switching to the appropriate rhombohedral cell (as shown in the figure) and attributing f_{Sc} scattering to the origin and $(1 - 2x) f_{Sc}$ scattering to the position at $\frac{1}{2}, \frac{1}{2}, \frac{1}{2}$. We will ignore the sulfur scattering because it does not contribute to the effect to be described. The Sc contribution to the structure factor is

$$F_{Sc} = f_{Sc} + (1 - 2x) f_{Sc} \exp -\pi i(h + k + l) . \tag{VIII.35}$$

If h + k + l (in the rhombohedral cell!) is even (e.g., for 222) then $F_{Sc} = f_{Sc} + (1 - 2x) f_{Sc}$. If h + k + l is odd (e.g., for 111) then $F_{Sc} = f_{Sc} - (1 - 2x) f_{Sc} = xf_{Sc}$, and thus h + k + l odd yields a weak reflection which vanishes with x. It is desirable to find the cubic (NaCl-type) indexing for these reflections.

The rhombohedral and cubic lattice vectors are related by $\mathbf{a}_{rh} = \mathbf{a}_c + \frac{1}{2}(\mathbf{b}_c + \mathbf{c}_c)$, $\mathbf{b}_{rh} = \mathbf{b}_c + \frac{1}{2}(\mathbf{a}_c + \mathbf{c}_c)$ and $\mathbf{c}_{rh} = \mathbf{c}_c + \frac{1}{2}(\mathbf{a}_c + \mathbf{b}_c)$ (see Fig. VIII.8) or

$$\begin{pmatrix} 1 & \frac{1}{2} & \frac{1}{2} \\ \frac{1}{2} & 1 & \frac{1}{2} \\ \frac{1}{2} & \frac{1}{2} & 1 \end{pmatrix} \begin{pmatrix} \mathbf{a}_c \\ \mathbf{b}_c \\ \mathbf{c}_c \end{pmatrix} = \begin{pmatrix} \mathbf{a}_{rh} \\ \mathbf{b}_{rh} \\ \mathbf{c}_{rh} \end{pmatrix} \tag{VIII.36}$$

To find the inverse matrix set

$$\begin{pmatrix} 1 & \frac{1}{2} & \frac{1}{2} \\ \frac{1}{2} & 1 & \frac{1}{2} \\ \frac{1}{2} & \frac{1}{2} & 1 \end{pmatrix} \begin{pmatrix} a & b & c \\ d & e & f \\ g & h & i \end{pmatrix} = \begin{pmatrix} 1 & 0 & 0 \\ 0 & 1 & 0 \\ 0 & 0 & 1 \end{pmatrix} \tag{VIII.37}$$

and solve for a through i:

$$\begin{pmatrix} \frac{3}{2} & -\frac{1}{2} & -\frac{1}{2} \\ -\frac{1}{2} & \frac{3}{2} & -\frac{1}{2} \\ -\frac{1}{2} & -\frac{1}{2} & \frac{3}{2} \end{pmatrix} \begin{pmatrix} \mathbf{a}_{rh} \\ \mathbf{b}_{rh} \\ \mathbf{c}_{rh} \end{pmatrix} = \begin{pmatrix} \mathbf{a}_c \\ \mathbf{b}_c \\ \mathbf{c}_c \end{pmatrix} . \tag{VIII.38}$$

It turns out that the matrix for converting hkl values is the same as that for converting **a**, **b**, **c**'s, i.e.,

$$\begin{pmatrix} \frac{3}{2} & -\frac{1}{2} & -\frac{1}{2} \\ -\frac{1}{2} & \frac{3}{2} & -\frac{1}{2} \\ -\frac{1}{2} & \frac{1}{2} & \frac{3}{2} \end{pmatrix} \begin{pmatrix} h \\ k \\ l \end{pmatrix}_{rh} = \begin{pmatrix} h \\ k \\ l \end{pmatrix}_{c}, \tag{VIII.39}$$

accordingly

$$\begin{pmatrix} \frac{3}{2} & -\frac{1}{2} & -\frac{1}{2} \\ -\frac{1}{2} & \frac{3}{2} & -\frac{1}{2} \\ -\frac{1}{2} & -\frac{1}{2} & \frac{3}{2} \end{pmatrix} \begin{pmatrix} 1 \\ 1 \\ 1 \end{pmatrix} = \begin{pmatrix} \frac{1}{2} \\ \frac{1}{2} \\ \frac{1}{2} \end{pmatrix}. \tag{VIII.40}$$

The significance of this result (because the indices in the cubic cell are half-integral) is that the rhombohedral 1, 1, 1 reflection has no counterpart in the NaCl-type case. As we have seen, this reflection vanishes for $x = 0$, or for equal $(1 - x)$ occupancy at both 0, 0, 0 and $\frac{1}{2}, \frac{1}{2}, \frac{1}{2}$ in the rhombohedral cell. Such reflections are called superstructure reflections. They derive intensity from the fact that two parts of the structure (in this case alternate Sc containing planes along the cubic 1, 1, 1 direction) are incompletely destructively interfering, and they vanish when disorder causes the two parts to scatter equally. The 2, 2, 2 reflection in the rhombohedral case is the cubic 1, 1, 1 reflection and this "substructure" reflection is present for both the ordered and the disordered structure alike.

If there is an odd number of odd indices in hkl for the rhombohedral case, then $h + k + l$ is odd, and $F_{Sc} = xf_{Sc}$. Furthermore in such cases the so-called hkl values for the cubic case will be half integral and thus show that such a reflection is from the "superstructure" on the NaCl-type lattice.

VIII.7 Incommensurate Ordered Superstructure

The order-disorder transition for $Sc_{1-x}S$ discussed above occurs for $Sc_{0.8}S$ at about 700 °C. In the ordered structure (which is actually only partially ordered) the vacancies are located in alternate Sc containing planes along the 1, 1, 1 direction, but are distributed randomly within these planes (which are then 60% occupied). The vacancies within these planes order further at lower temperatures [43]. The occurrence of diffraction satellites (superstructure spots) at ± 0.349 of the 402 diffraction maximum is observed in the electron diffraction pattern.

It is a valuable exercise to consider the origin of such "incommensurate" peaks.

First consider a (hypothetical) commensurate ordering that results when the vacancies accumulate in every third 4, 2, 0 plane. Such ordering would yield a commensurate superstructure, i.e., the superstructure diffraction peaks would occur at 0.333... of the 4, 2, 0 spacing, rather that at the observed 0.346. However, if occasional faults with vacancies in alternate planes, rather than every third plane, occur in a more-or-less systematic fashion, then the fractional occupancy (represented by φ) can be plotted versus the direction perpendicular to the 4, 2, 0 planes, and a population wave, $\varphi(x)$, represented as a square wave (Fig. VIII.9) results. The vertical lines represent the 4, 2, 0 planes which are labeled by integers along the horizontal axis. Note that $\varphi(x)$ is incommensurate with the lattice periodicity.

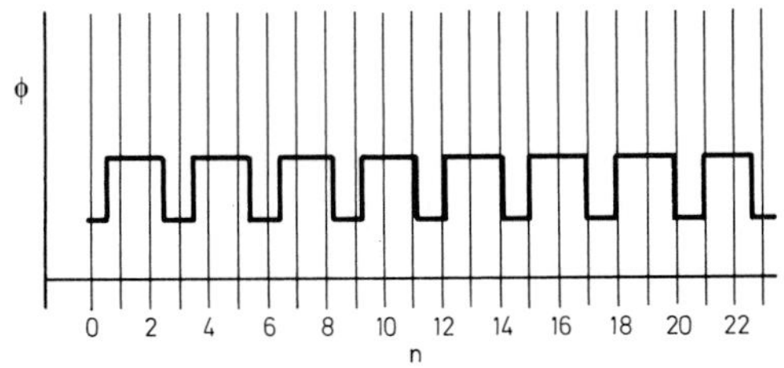

Fig. VIII.9. A population wave (φ = fractional occupancy of sites along the $2\bar{2}0$ direction) which is incommensurate with the lattice periodicity (vertical lines represent $2\bar{2}0$ planes).

The expansion of a square wave in a Fourier series is a textbook example of the application of Fourier series:

$$\phi(x) = \frac{\phi_1 + 2\phi_2}{3} + \sum_{m=0} \frac{2(\phi_1 - \phi_2)}{m\pi} L \sin \frac{m\pi}{3} \cos 2m\pi \frac{x}{L} . \qquad \text{(VIII.41)}$$

The fractional occupancy of the n'th plane is $\varphi(n)$, hence

$$F = \sum_{n=1}^{N} \phi(n) f_{Sc} \exp - 2\pi \text{ inh} \qquad \text{(VIII.42)}$$

and substitution of the Fourier expansion of $\varphi(X)$ evaluated at $x = n$ with $\cos 2m\pi^{n}/L$ replaced by $(\exp 2\pi inm/_{L} + \exp -2\pi inm/_{L})$ yields

$$F = \sum_{m=1}^{N} \left(\frac{\phi_1 + 2\phi_2}{3}\right) f_{Sc} \exp - 2\pi \text{ inh} + \sum_{m=0}^{N} \left(\frac{2\phi_1 - \phi_2}{m\pi}\right) L f_{Sc} \sin \frac{m\pi}{3}$$

$$\times \sum_{n=1}^{N} \left[\exp - 2\pi \text{ in} \left(h + \frac{m}{L}\right) + \exp - 2\pi \text{ in} \left(h - \frac{m}{L}\right)\right]. \qquad \text{(VIII.43)}$$

According to the result obtained at the start of this chapter, that very large sums of complex exponentials in which the exponents (phase angles) are incremented in the summation have significant magnitude only when the increment is very close to 2π, we find from this expression that diffraction intensity is expected at h = integer with the effective scattering factor of the overall average occupancy

$\frac{\varphi_1 + 2\varphi_2}{3}$, i.e., substructure diffraction, and incommensurate superstructure diffraction is expected at integral values of $h \pm m/L$ with $m =$ integer (in the case under discussion only $m = 1$ is observed) and $L =$ the wave length of the population wave. Since superstructure diffraction intensity is observed at $h \pm 0{,}346$ it is concluded that $L = 2.89$ is the wave length of the population wave in units of interplanar (4, 2, 0) spacing.

Several features of the diffraction effects should be noted. First, the substructure reflections are unaffected by the ordering. Note in particular that when m is a multiple of 3, so that the superstructure reflections would approximately coincide with the substructure reflections, then the $\sin \frac{m\pi}{3}$ term in the superstructure term causes the whole term to vanish. Second, the superstructure term vanishes when $2\varphi_1 = \varphi_2$, as should be the case. Finally, the relationship of the discussion here (incommensurate ordering) to that of the previous section (commensurate ordering) is made clear when it is realized that L could be an integer and a commensurate superstructure would result. In this case it would be possible either to determine the new space lattice and describe the diffraction effects in terms of that lattice, as was done in the previous section, or to consider the diffraction in terms of submultiples (m/L) as was done here. According to the results of Sect. VII.6, the former is appropriate when the **K** point implies the absence of vector invariants, otherwise the latter is appropriate.

VIII.8 Electron Density and Diffraction

The reasoning used to develop the x-ray structure factor as a sum over atomic positions can be generalized to the case of a continuous electron distribution with the electron density function times the volume element, $\rho(\mathbf{r})\,dV$, taking the place of f_j (since the number of electrons at $\mathbf{r}$ is a measure of the scattering efficiency at $\mathbf{r}$) and the sum replaced by an integral, i.e.,

$$F_{h'} = \int_{\substack{\text{unit}\\\text{cell}}} \varrho(\mathbf{r}) \exp - 2\pi\, i\mathbf{K}_{h'} \cdot \mathbf{r}\, dV \tag{VIII.44}$$

where h labels the reciprocal vector, and hence the diffracting family of planes.

Since the electron density distribution is periodic it can be expanded in a Fourier series

$$\varrho(\mathbf{r}) = \sum_h C(h) \exp 2\pi i\mathbf{K}_h \cdot \mathbf{r} \tag{VIII.45}$$

which, upon substitution in F_n, gives

$$F_{h'} = \sum_h C(h) \int_{\substack{\text{unit}\\\text{cell}}} \exp 2\pi i\mathbf{r} \cdot (\mathbf{K}_h - \mathbf{K}_{h'})\, dV\,. \tag{VIII.46}$$

Since $\mathbf{K}_h - \mathbf{K}_{h'} = \mathbf{K}_{h''}$, a reciprocal lattice vector, and in general $\mathbf{r} \cdot \mathbf{K}_{h''} = hx + ky$

+ pz, the integrand can be factored into the product of three integrals each of which, unless h = 0, is of the form

$$\int_{x=0}^{x=1} [\exp(2\pi ihx)]\, dx = \int_{x=0}^{x=1} (\cos 2\pi hx)\, dx + i \int_{x=0}^{x=1} (\sin 2\pi hx)\, dx \quad \text{(VIII.47)}$$

$$= \frac{1}{2\pi k} \left[\sin 2\pi hx \Big|_{x=0}^{x=1} - i \cos 2\pi hx \Big|_{x=0}^{x=1}\right] \quad \text{(VIII.48)}$$

$$= 0 \quad \text{(VIII.49)}$$

Thus the integral vanishes unless h = k = 1 = 0, i.e., unless $\mathbf{K}_h = \mathbf{K}_{h'}$, in which case

$$\int_{\substack{\text{unit}\\\text{cell}}} \exp 2\pi i \mathbf{K}_{h''} \cdot \mathbf{r}\, dV = V_{cell} \,. \quad \text{(VIII.50)}$$

What has been demonstrated is that the complex exponentials $\exp 2\pi i\, \mathbf{K} \cdot \mathbf{r}$ are orthogonal and normalizable. It follows that out of the sum on the right hand side of Eq. VIII.46 the only term that is nonzero is $C(h')\, V_{cell}$ and $C(h') = F_h/V_{cell}$. It follows that

$$\varrho(\mathbf{r}) = \frac{1}{V_{cell}} \sum_h F_h \exp 2\pi i \mathbf{K}_h \cdot \mathbf{r} \,. \quad \text{(VIII.51)}$$

Since the magnitude of F_h is obtained directly from diffraction intensities, the structure ($p(\mathbf{r})$) can be determined when phase angles are determined. The process of structure determination from diffraction data is the problem of determining phase angles consistent with the determined intensities and other known features of the solid (elemental content, minimum interatomic distances, density, etc.)

In the case of centrosymmetric crystals, for each term in $F_h = \Sigma f_j \exp -2\pi i \mathbf{K}_h \cdot r_j$ of the form $f_j \exp -2\pi i \mathbf{K}_h \cdot \mathbf{r}_j$ there is another for $-\mathbf{r}_j$, i.e., $f_j \exp 2\pi i \mathbf{k}_h \cdot \mathbf{r}_j$ and the sum of these is real, namely $2f_j \cos 2\pi \mathbf{K} \cdot \mathbf{r}_j$. Therefore for centrosymmetric crystals

$$F_h = 2 \sum_j f_j \cos 2\pi \mathbf{K}_h \cdot \mathbf{r}_j \quad \text{(VIII.52)}$$

where the sum is over half the atom sites and the other half are related to these by an inversion through the origin. Since this F_h is real, its phase angle is either 0 or π, i.e., F_h might be positive or negative. Hence, in the case of centrosymmetric crystals the phase problem is one of assigning the appropriate sign to $|F_h|$ determined from the diffraction intensity.

VIII.9 Problems

1. Find the extinction rules for body-centered and end-centered lattices.
2. Find the extinction rules for a c glide perpendicular to **b**.
3. Find the extinctions for Cmcm.
4. The vacancies on the metal atom positions of the NaCl-type structure with a = 5,17 Å order to form a C2/m lattice with a = 12.11, b = 3.655 and c = 6.33 Å and $\beta = 100°$. Determine the monoclinic h, k, l values for the two lowest angle superstructure reflections. What are the Bragg angles of these reflections (Cu radiation, $\lambda = 1.54$ Å)?
5. Describe the difference between the diffraction of NbO (described in Chap. 3) and that of hypothetical NbO with the NaCl-type structure.
6. Describe how you would differentiate between a solid MY with the cubic ZnS-type structure and one with the Y atoms distributed over all the tetrahedral sites of the CaF_2 structure. Can such a transition occur continuously? What would be the changes in the X-ray powder diffraction pattern with temperature if the change occurred continuously?
7. It is generally the case that superstructure reflections are substantially weaker than substructure reflections. In $Sc_{1-x}S$ with the vacancies ordered in a alternate metal containing planes along 1, 1, 1 it is observed that the rhombohedral 1, 1, 1 reflections (which are superstructure reflections) are stronger than the rhombohedral 2, 2, 2 (substructure) reflections. Explain.
8. Consider a hypothetical process which carries a NiAs-type solid to a WC-type solid through a disordered intermediate solid. Contrast qualitatively the X-ray diffraction patterns of the 3 solids, WC, NiAs and disordered intermediate, with emphasis upon 00I reflections. A very brief answer is requested, since working out the diffraction of disordered solids in detail is laborious and time consuming.

Chapter IX

Order-Disorder Transition and Disordered Structures

IX.1 Introduction

This chapter consists of two parts. The first part treats applications of Landau's theory of symmetry and phase transitions as discussed in Chap. VII to a variety of order-disorder processes that have been mentioned in earlier chapters, namely: CsCl-type to bcc-type, $CdCl_2$-type to NaCl-type, incommensurate ordering in $Sc_{1-x}S$, and CdI_2-type to NiAs-type. The second part is a discussion of a few known ordered defect structures which provide examples of interesting solid-state ordering phenomena that are at this time not well understood. These structures include $Ti_{1-x}O_{1-x}$, Lu_3S_4, and the defect titanium and chromium sulfides and $WO_{2.41}$. There are many other cases of ordered defect structures, however, these examples are sufficient to convey the point that there are some rather complicated structural effects accompanying the ordering of vacancies in transition metal compounds.

IX.2 CsCl-Type to bcc-Type

The classic example of an order-disorder transition is the CsCl-type to bcc-type transition in CuZn discussed in a general way in Chapter IV and in greater detail in Chap. VII. The symmetry operations of bcc are those of O_h, and those of O_h with the translations, including $(\mathbf{a} + \mathbf{b} + \mathbf{c})/2$. For the CsCl-type structure all operations of O_h with a translational component $(\mathbf{a} + \mathbf{b} + \mathbf{c})/2$ are absent, otherwise the operations are the same as for the bcc-type structure. In other words, the space-group of CsCl-type is a subgroup of that of bcc-type, and the symmetry operations lost correspond to a single irr. rep., namely one with a basis function which is symmetric with respect to all operations of O_h (permute x, y, z with changes in signs) but antisymmetric with respect to transition by $(\mathbf{a} + \mathbf{b} + \mathbf{c})/2$. An example of a function with this symmetry is

$$\varphi = \cos 2\pi x + \cos 2\pi y + \cos 2\pi z \,. \qquad \text{(IX.1)}$$

The transition corresponds to totally symmetric small representation at the $\mathbf{k}$ point $\mathbf{a}^* + \mathbf{b}^* + \mathbf{c}^*$. This $\mathbf{k}$ vector is carried into itself under all symmetry operations of O_h, and thus the irr. rep. of the space group is one dimensional. Since $\varphi \rightarrow -\varphi$ under translation by $(\mathbf{a} + \mathbf{b} + \mathbf{c})/2$, there can be no third-order invariant in the expansion of G, and the transition from CsCl-type to bcc-type

(Pm3m to Im3m) meets the first three conditions of the Landau theory. Also $g^0(\mathbf{k})$ includes sufficiently many symmetry operations (including inversion) to assure that the 4th condition is met as well, and thus a continuous (second-order) transition is allowed for $\beta \rightarrow \beta'$ brass by the Landau theory of symmetry and phase transitions.

IX.3 $CdCl_2$-Type to NaCl-Type

The phase transition in $Sc_{1-x}S$ mentioned in Chap. IV and used again in Chapter VIII for the purpose of discussing diffraction effects accompanying a commensurate ordering transition appears to occur as a second-order phase transition.

As discussed in Chap. IV, stoichiometric ScS has the NaCl-type structure. At high temperatures the compound becomes nonstoichiometric with Sc vacancies. When the sample has been cooled slowly the vacancies are preferentially located in alternate Sc containing planes along the [1, 1, 1] direction (Fig. VIII.8). Fm3m symmetry elements that remain after ordering are the identity, the 3-fold rotations along [1, 1, 1], the two-folds perpendicular to [1, 1, 1], the inversion and the product of all the aforementioned with the inversion, i.e., the operations that remain are

$$\varepsilon,\ C_3,\ C_3^2,\ C_{2(y-x)},\ C_{2(z-y)},\ C_{2(z-x)},\ i,\ \bar{C}_3,\ \bar{C}_3^5,\ \sigma_{(y-x)},\ \sigma_{(z-y)},\ \sigma_{(z-x)},$$

which are the rotational symmetry elements of the space group R$\bar{3}$m. The primitive translations of the ordered structure are

$$\mathbf{a}_{rh} = \mathbf{a}_c + \frac{\mathbf{b}_c + \mathbf{c}_c}{2} \tag{IX.2}$$

$$\mathbf{b}_{rh} = \mathbf{b}_c + \frac{\mathbf{a}_c + \mathbf{c}_c}{2} \tag{IX.3}$$

and

$$\mathbf{c}_{rh} = \mathbf{c}_c + \frac{\mathbf{a}_c + \mathbf{b}_c}{2}, \tag{IX.4}$$

and thus the translations are a subgroup of the translations of the cubic, disordered structure. Thus the group-subgroup condition of Landau is met by this order-disorder transition.

Next we examine whether the distortion corresponds to a single irr. rep. Since all of the translational symmetry operations that are lost obey

$$(\mathbf{a}^* + \mathbf{b}^* + \mathbf{c}^*) \cdot \mathbf{T} = n/2\,, \quad n = \text{odd integer} \tag{IX.5}$$

and all that remain obey

$$(\mathbf{a}^* + \mathbf{b}^* + \mathbf{c}^*) \cdot \mathbf{T} = n\,, \quad n = \text{integer} \tag{IX.6}$$

it follows that the distortion "corresponds to" the **k** vector $(\mathbf{a}^* + \mathbf{b}^* + \mathbf{c}^*)/2$, i.e., $\cos(2\pi(\mathbf{a}^* + \mathbf{b}^* + \mathbf{c}^*)/2 \cdot \mathbf{r}) = \cos \pi(x + y + z)$ is a basis function for the irr. rep. that is symmetric with respect to translations that remain and antisymmetric with respect to those that are lost. Furthermore all of the rotational operations listed above are in the point group of $(\mathbf{a}^* + \mathbf{b}^* + \mathbf{c}^*)/2$, thus the ordering corresponds to the totally symmetric small representation of Fm3m at $\mathbf{k} = \frac{1}{2}, \frac{1}{2}, \frac{1}{2}$, and the symmetry change does correspond to a single irr. rep.

In order to consider the invariants of various order we must determine what **k** vectors are in the star. The symmetry operations of Fm3m carry $\frac{1}{2}, \frac{1}{2}, \frac{1}{2}$ into $-\frac{1}{2}, \frac{1}{2}, \frac{1}{2}$; $\frac{1}{2}, -\frac{1}{2}, \frac{1}{2}$; and $\frac{1}{2}, \frac{1}{2}, -\frac{1}{2}$ as well as the four vectors that result from these by inversion. Each **k** vector is related to its inverse by 1, 1, 1 and thus $\mathbf{k} = \mathbf{ik} + \mathbf{K}$. Note that although $\frac{1}{2}, \frac{1}{2}, \frac{1}{2}$ and $-\frac{1}{2}, \frac{1}{2}, \frac{1}{2}$ differ by 1, 0, 0 they are not the same modulo a reciprocal lattice vector because 1, 0, 0 is not a reciprocal lattice vector in the case of a face-centered real lattice. There are four basis functions which form a basis for the irr. rep. to which the transition corresponds, e.g.:

$$\varphi_1 = \cos \pi(x + y + z) \tag{IX.7}$$

$$\varphi_2 = \cos \pi(-x + y + z) \tag{IX.8}$$

$$\varphi_3 = \cos \pi(x - y + z) \tag{IX.9}$$

$$\varphi_4 = \cos \pi(x + y - z)\,, \tag{IX.10}$$

and the rotational symmetry operations carry these functions into each other without sign change, while the translations carry them into themselves (if the translations remain) or into their negative (if the translations are lost).

It follows that in the expansion for G:

$$G = G^0 + A\eta^2 + Bf^{\langle 3\rangle}(\gamma_i)\eta^3 + Cf^{\langle 4\rangle}(\gamma_i)\eta^4 + \ldots \tag{IX.11}$$

based upon a generalized electron density corresponding to this irr. rep.:

$$\varrho^{\text{distorted}} = \varrho^{\text{sym}} + [\gamma_1\,\Delta\varrho_1 + \gamma_2\,\Delta\varrho_2 + \gamma_3\,\Delta\varrho_3 + \gamma_4\,\Delta\varrho_4]\,\eta\,, \tag{IX.12}$$

there are no third-order invariants and the fourth-order term looks like:

$$C_1 \sum \gamma_i^4 + C_2 \sum_{i \neq j} \gamma_i^2\gamma_j^2 + C_3\gamma_1\gamma_2\gamma_3\gamma_4\,. \tag{IX.13}$$

The stable distortion will correspond to minima of this function subject to the restraint $\Sigma\,\gamma_i^2 = 1$. Employing Lagrange's method of undetermined multipliers with λ as the multiplier and minimization with respect to γ_1, γ_2, γ_3 and γ_4 yields:

$$4C_1\gamma_1^3 + 2C_2\gamma_1(\gamma_2^2 + \gamma_3^2 + \gamma_4^2) + C_3\gamma_2\gamma_3\gamma_4 + 2\lambda\gamma_1 = 0 \tag{IX.14}$$

$$4C_1\gamma_2^3 + 2C_2\gamma_2(\gamma_1^2 + \gamma_3^2 + \gamma_4^2) + C_3\gamma_1\gamma_3\gamma_4 + 2\lambda\gamma_2 = 0 \quad \text{(IX.15)}$$

$$4C_1\gamma_3^3 + 2C_2\gamma_3(\gamma_1^2 + \gamma_2^2 + \gamma_4^2) + C_3\gamma_1\gamma_2\gamma_4 + 2\lambda\gamma_3 = 0 \quad \text{(IX.16)}$$

$$4C_1\gamma_4^3 + 2C_2\gamma_4(\gamma_1^2 + \gamma_2^2 + \gamma_3^2) + C_3\gamma_1\gamma_2\gamma_3 + 2\lambda\gamma_4 = 0 \quad \text{(IX.17)}$$

for which there are three types of solutions, $\gamma_1 = 1$, $\gamma_2 = \gamma_3 = \gamma_4 = 0$ (the $R\bar{3}m$ ordering observed), $\gamma_1 = \gamma_2 = \gamma_3 = \gamma_4 = \frac{1}{2}$ (a f.c.c. ordering with a $2a_{NaCl}$), and $\gamma_1 = \gamma_2 = \gamma_3 = \frac{1}{2}$, $\gamma_4 = -\frac{1}{2}$ an ordering leading to Fd3m symmetry.

IX.4 Incommensurate Ordering in $Sc_{1-x}S$

The incommensurate ordering that occurs in the ordered $R\bar{3}m$ structure at yet lower temperatures (discussed in Chap. VIII) occurs at a $\mathbf{k} = \mu(\mathbf{a}^* - \mathbf{c}^*)$ point of the reciprocal space of $R\bar{3}m$. The point group of this **k** contains only a two-fold rotation and the identity. However, the Landau 4th condition (no vector invariants of **k**) is appropriate only for transitions to a given fixed point in reciprocal space. Since the transition in question here is to **k** vectors which can change with thermodynamic state (variable μ) it is of the type for which $\alpha(T, P, X) = 0$ rather than $\boldsymbol{\alpha} \equiv 0$ by symmetry. The incommensurate nature of the resultant structure is a natural consequence of this fact. That is, the commensurate transition which exactly triples the periodicity in the cubic 4, 2, 0 direction (L = 3 in Eq. VIII.39) has no special symmetry reason to occur, since the absence of a vector invariant at **k** with $\mu = \frac{1}{3}$ means that a minimum in G will occur for this **k** only because $\boldsymbol{\alpha}(T, P, X)$, which varies with the state variables, happens to vanish at some state points.

IX.5 CdI_2-Type to NiAs-Type

In $Cr_{1-x}S$ at high temperatures there occurs a second-order disordering of the vacancies which initially are located in alternate Cr containing planes along **c** (CdI_2 structure type, space group $P\bar{3}m1$) to produce equal occupation of all Cr sites (NiAs structure type, space group $P6_3/mmc$) [36]. The transition occurs with no change in translational periodicity (at $\mathbf{k} = 0$) but with a loss of all essential symmetry elements with translational part 0, 0, $\frac{1}{2}$ (i.e., every other symmetry element listed in Table VII.1). Thus the ordering obeys the group-subgroup relation, and belongs to a single irr. rep. A basis function for the one-dimensional irr. rep. is $\cos 2\pi z$, which is carried into itself by all operations of $P6_3/mmc$ except those which increase z by $\frac{1}{2}$, in which case it is carried into its negative. Clearly $\varphi^3 \rightarrow -\varphi^3$ under all symmetry operations that increase z by $\frac{1}{2}$ and thus there are no third-order invariants, and, since $g^0(k)$ is D_{6h}, Landau's 4th condition is met. Thus the transition can occur as a continuous process.

IX.6 Vacancy Ordering in a Few Defect NaCl-Type Solids

It was noted in Sect. IX.4 that the lower temperature ordering in $Sc_{1-x}S$ was very nearly a tripling of the periodicity in the 420 direction. The purpose of this section is to show that the 420 planes appear to play an important role in vacancy ordering in defect rock salt materials. In the case of stoichiometric TiO ($Ti_{5/6}O_{5/6}$) with vacancies on both the Ti and O sublattices, the vacancies order to form a monoclinic (A2/m, a = 585.5, b = 934, c = 414 pm, $\gamma = 107° 32'$) structure in which the vacancies are in every sixth 420 plane (Fig. IX.1).

The compound Sc_2S_3 (Fddd, a = 1041, b = 2205, c = 738 pm) [44] shown in Fig. IX.2 forms with the atom positions essentially unchanged from those in NaCl-type $Sc_{1-x}S$, but with the vacancies ordered as shown. The order-disorder transition is first order in this case. In this structure each sixth cubic 420 plane is equivalent, and the vacancy ordering doubles the periodicity in the c direction. Thus the ordering corresponds to $\mathbf{k} = \frac{2\mathbf{a}}{3} + \frac{\mathbf{b}^*}{3} + \frac{\mathbf{c}^*}{2}$. However, as can be seen in the figure, the vacancies are aggregated in adjacent pairs of 420 lines to yield the overall stoichiometry ($\frac{2}{6} = \frac{1}{3}$ of the metal sites vacant).

The structure of Lu_3S_4 (Fddd, a = 1074.7, b = 2281.3, c = 760.2 pm) [45] is shown in Fig. IX.3. The occurrence in this structure of fractional occupancy population waves is unusual. The site occupation in metal containing planes varies periodically ... 54%, 75%, 83%, 84%, 83%, 75%, 54%, ... along the cubic

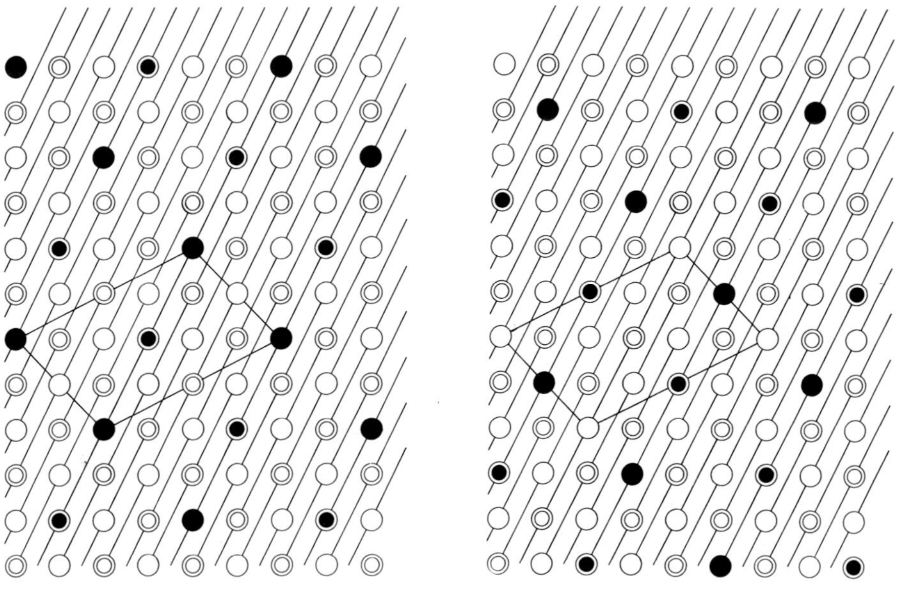

Fig. IX.1. The TiO structure. Layers seperated by $\frac{1}{2}$ along the unique axis shown. Single circles represent Ti positions in one level. O positions in the next. Double circles represent oxygen atoms in the level in which titanium atoms are represented by single circles, and vice versa. Blackened circles represent vacant positions.

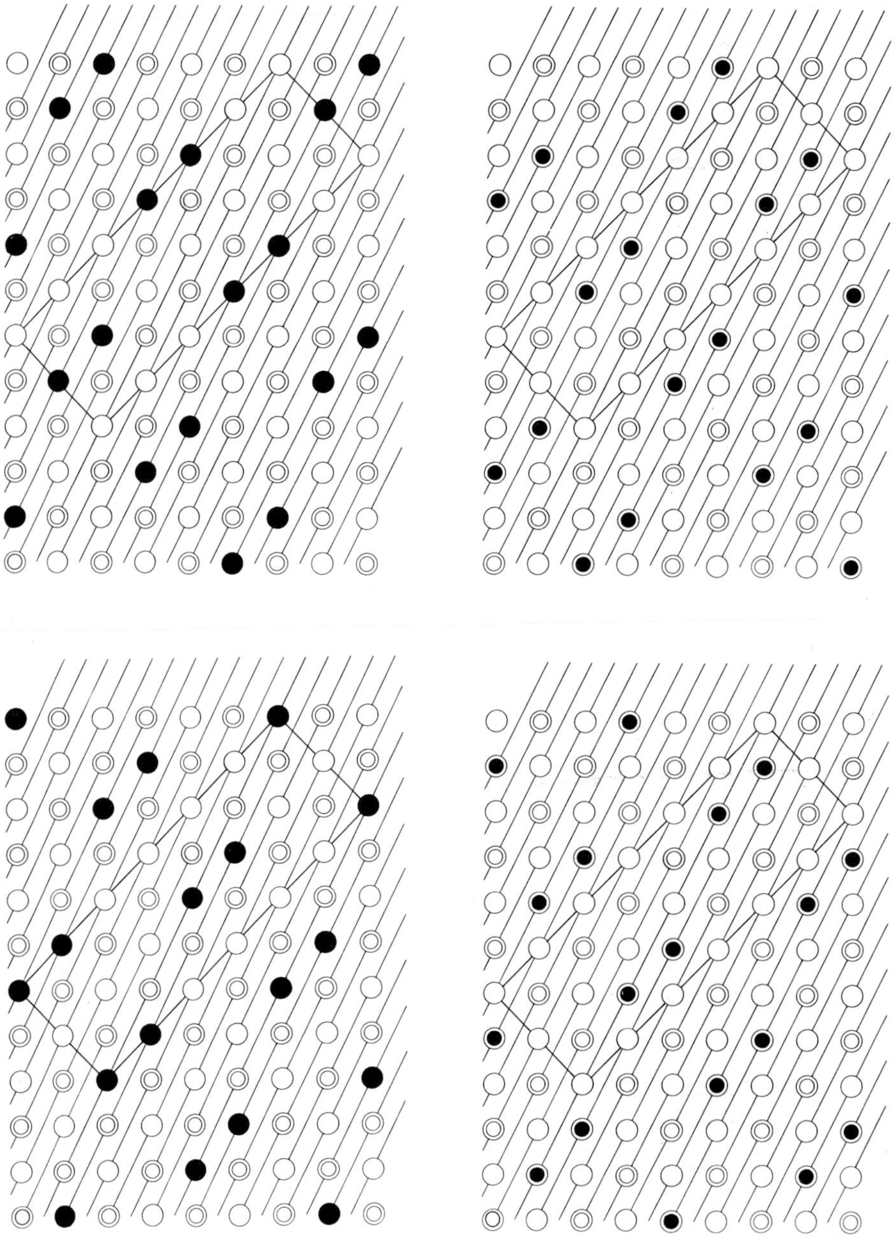

Fig. IX.2. The Sc and S positions in layers along the c axis in Sc_2S_3. The Sc atoms are in positions with single circles in one layer and with double circles in the next layer, etc. Vacant positions are identified by darkening.

420 direction. The periodicity along the c direction is doubled by the occurrence along c of pairs of adjacent NaCl-type 002 planes with the same origin for the population wave along the cubic 420 direction, but with the origin alternating by three 420 planes (population wave exactly out of phase) from pair to adjacent pair. It is interesting to note that the ordering, as for Sc_2S_3, corresponds to $\mathbf{k} = \frac{2\mathbf{a}^*}{3} + \frac{\mathbf{b}^*}{3} + \frac{\mathbf{c}^*}{2}$. This $\mathbf{k}$ vector is of the type $\frac{\mathbf{a}^* + \mathbf{b}^* + \mathbf{c}^*}{2} + \frac{\mathbf{a}^* - \mathbf{b}^*}{\lambda}$, with $\lambda = 6$. The group of the wave vector is C_2, suggesting that there are some incommensurate components to the structure owing to the existence of vector invariants at $\mathbf{k}$ (see Sect. VII.6), i.e., there is no symmetry reason why λ should be exactly six.

IX.7 Vacancy Ordering in Defects $Cr_{1-x}S$

A schematic phase diagram for the solid sulfides of chromium is shown in Fig. IX.4 [46]. The wide homogeneity range regions above 550 K and 570 K represent the NiAs-type and CdI_2-type structures, with the transition between them changing from second-order at higher temperatures (dashed line) to first-order at lower temperatures (two-phase region). At yet lower temperatures these solid solution regions disproportionate into line compounds Cr_5S_6, Cr_7S_8, and CrS. The two disproportionation reactions that occur with decreasing temperature are:

$$Cr_{1-x}S\ (CdI_2\text{-type}) = Cr_5S_6 + Cr_7S_8 \tag{IX.18}$$

and

$$Cr_{1-x}S\ (\text{NiAs-type}) = Cr_7S_8 + CrS\,. \tag{IX.19}$$

The discussion of Sect. V.9 relates to these reactions, and leads directly to the conclusion that these reactions occur exothermically as written, or that the solid solution compounds on the left-hand side of the reactions are formed endothermically and endoentropically from the stoichiometric solids on the right-hand sides. The general correlation of high configurational entropies with random defects is consistent with this observation, but probably not sufficient to quantitatively determine the ΔS values.

The nature of the vacancy ordering in the sulfides of chromium is of interest. In Cr_7S_8 the vacancies are in alternate planes along the c axis (as in the case for the CdI_2-type disordered structure) but the a axis is doubled and the c axis is tripled. In this enlarged cell the vacancies are located at the positions $\frac{1}{2}, \frac{1}{2}, 0$; $\frac{1}{2}, 0, \frac{1}{3}$; $0, \frac{1}{2}, \frac{2}{3}$, i.e., they form a spiral (space group $P3_1$) about the c axis.

In Cr_5S_6 [47] the vacancies order in a $\sqrt{3}a \times \sqrt{3}a$ superstructure (Fig. IX.5) in alternate Cr-containing planes, and the superstructure planes stack in a ABAB ... fashion in the $\sqrt{3}a \times \sqrt{3}a$ superstructure.

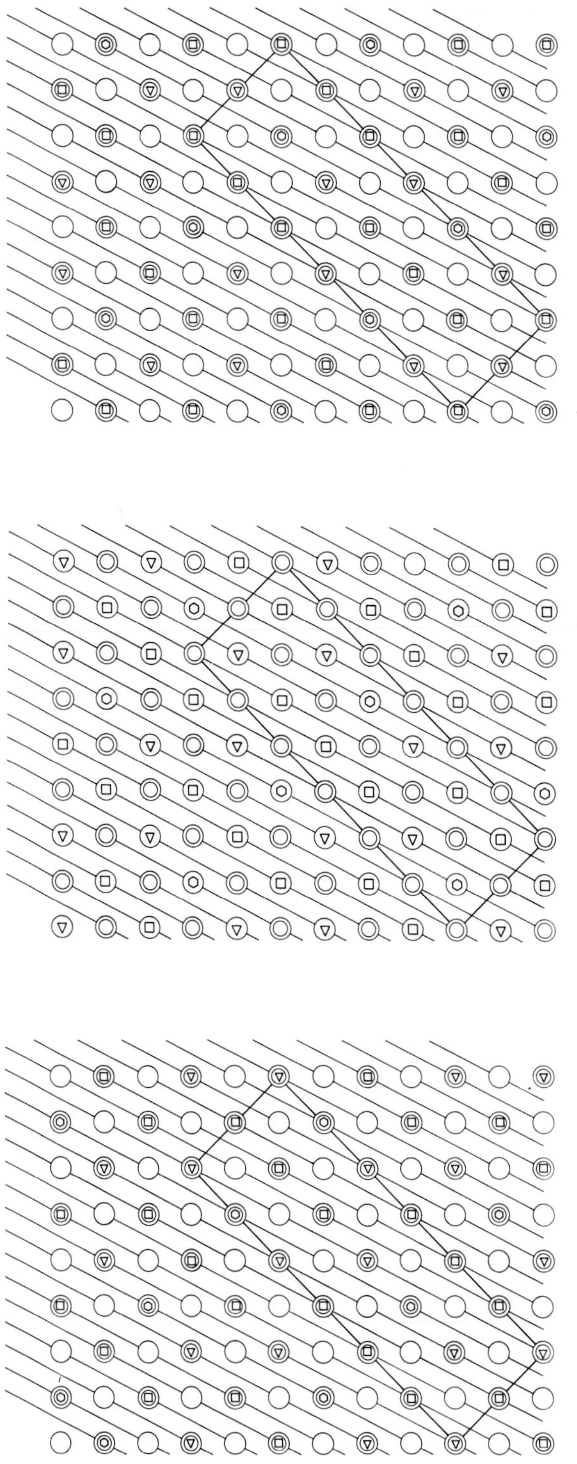

Fig. IX.3

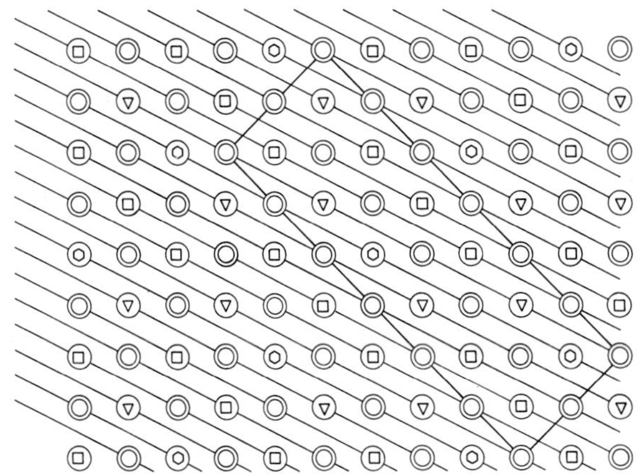

Fig. IX.3. The Lu_3S_4 structure. Layers separated by $\frac{1}{4}$ along the vertical axis. Positions marked with a square are 83.5% occupied, positions marked with a triangle are 75% occupied and positions marked with a hexagon are 54% occupied.

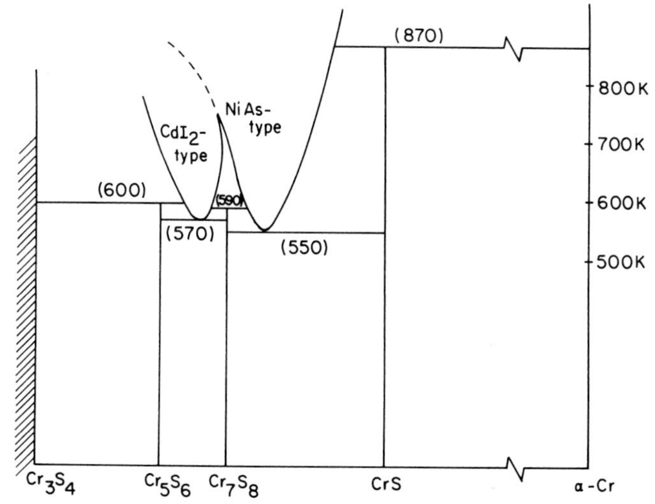

Fig. IX.4. The Cr—S phase diagram (Cr-rich side).

IX.8 The Ti_mS_n Intermediate Sulfides of Titanium

There are a number of intermediate titanium sulfides (between TiS and TiS_2) known, and they have been identified with the stoichiometries Ti/S $= \frac{8}{9}, \frac{4}{5}, \frac{3}{4}, \frac{2}{3}$ and $\frac{5}{8}$ [2]. To a first approximation (a number of recent papers have dealt with the subject of subtle alterations of some of these structures) the structures are as shown in Fig. IX.6. The numbers associated with the metal-atom layers in the figure are the fractional occupancies of the layers. Note that the stoichiometries

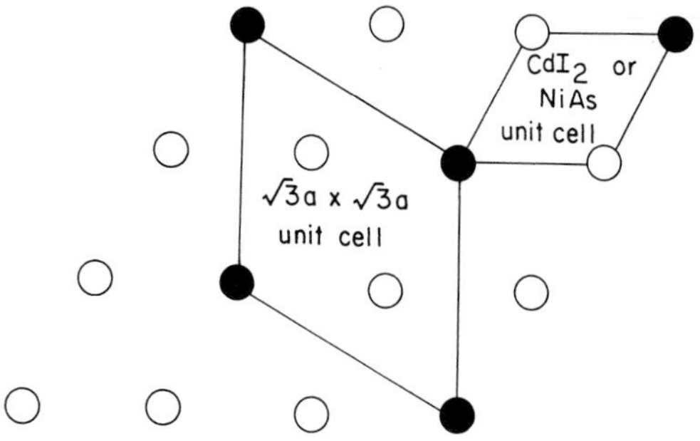

Fig. IX.5. The Cr—S superstructures.

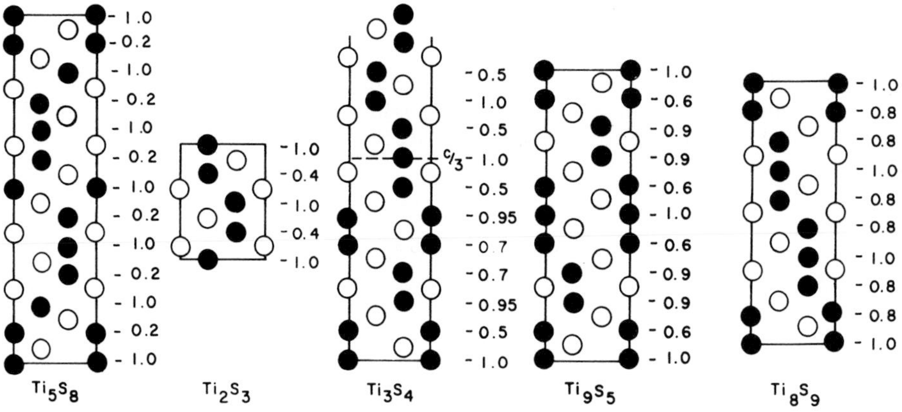

Fig. IX.6. The intermediate sulfides of titanium, Ti_nS_m with $1 < m/n < 2$.

given above are approximations to those observed. The Ti_mS_n structures can all be viewed as population wave structures (c.f., Lu_3S_4 and $Sc_{1-x}S$) with the waves running along the c direction, and as mixtures (on the unit cell scale) of defect NiAs-type and NaCl-type structures.

IX.9 Shear Structures

Referring back to Sect. IV.16 and the discussion of structures based upon ReO_3-type, it was mentioned there that defects resulting from modification of the ReO_3 structure could lead to nonstoichiometric, as well as stoichiometric solids. Consider Fig. IX.7 which shows an idealized representation of a crystallographic shear (CS) structure [48]. The outlined unit cell represents the idealized structure $W_{18}O_{52}$. If the long axis shown is considered, this axis is given by the relative displacement of the two parallel CS planes. The axis for the case shown, 8 square lengths to the right and 1 square length down (8,1) (one square length is the shortest oxygen-oxygen distance in the horizontal plane), has been observed. Two other such vectors observed by electron microscopy [48] are 9 square lengths to the

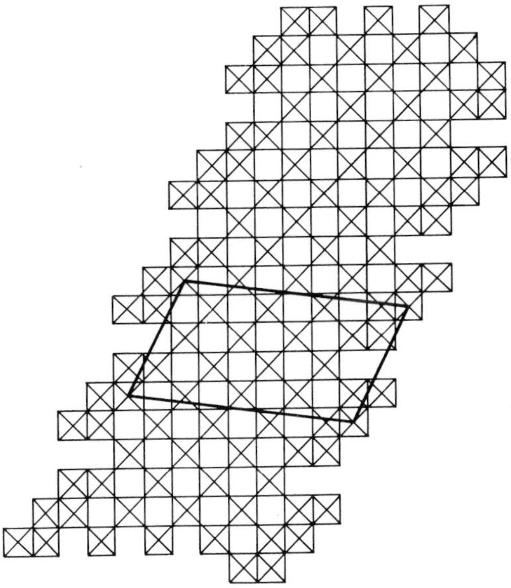

Fig. IX.7. The idealized $W_{18}O_{52}$ structure.

right (9, 0) and 7 to the right and two down (7, 2). The stoichiometries to which these idealized shear structures correspond are $W_{19}O_{55}$ (9, 0) and $W_{17}O_{49}$ (7, 2). All three of the stoichiometries are in the series W_nO_{3n-2}.

The observation that relates these idealized ordered structures to nonstoichiometry is that the three structures were observed to occur side by side in a more or less random fashion in needle shaped crystals with the stoichiometry $WO_{2.91}$. Thus, in spite of a substantial amount of short range order and long range order along the shear direction and along the vertical axis, there remained a nonstoichiometric feature, namely the variability of the **a** vector (7, 2), (8, 1) and (9, 0)) within the crystals providing, in principle, for nonstoichiometry in the range $W_{17}O_{49}$ to $W_{19}O_{55}$. This kind of nonstoichiometric feature appears to occur frequently in a variety of oxide structures with octahedral coordination.

IX.10 Summary

In this chapter and elsewhere throughout this text a variety of ordering phenomena which accompany the decrease in temperature of phases which have ranges of homogeneity at elevated temperatures has been discussed. Because of the diversity of those phenomena it is worthwhile to review and catagorize them. This section provides this review, and closes with a brief discussion of yet a different type of ordering phenomenon that requires an accompanying displacive transition. It should be mentioned that while short range ordering phenomena are known and discussed elsewhere, this chapter concentrates on the longer range phenomena such as give rise to diffraction effects of the type discussed in Chap. VIII.

In the beginning we assume that the sample has at high temperature an average structure typical examples of which are NaCl-type, NiAs-type or ReO_3-

type, at least in principle if not in fact. That is either we observe by x-ray diffraction techniques a structure with average fractional occupancy of, for example, the metal atom positions ($Sc_{1-x}S$ with the NaCl-type structure and (1 — x) Sc at 0, 0, 0, and 1S at 0, 0, $\frac{1}{2}$ or $Cr_{1-x}S$ with the NiAs-type structure and (1 — x) Cr at 0, 0, 0 and 1S at $\frac{1}{3}$, $\frac{2}{3}$, $\frac{1}{4}$, etc.), or we imagine that such a structure might exist if the sample were raised to a sufficiently high temperature and the vacancies disordered, but note that some other transformation (e.g., melting or phase transition) takes place, i.e., at temperatures of interest the hypothetical structure is unstable in the sense of Fig. V.5.

One common type of ordering is partial ordering such as is typified by both $Sc_{1-x}S$ in the NaCl-type structure and by $Cr_{1-x}S$ in the NiAs-type structure. In these cases, and a number of others, the metal vacancies tend to segregate into alternate metal containing planes. Such partial order is characterized by the fact that some sites continue to be "fractionally occupied" even though a superstructure is formed, and includes cases such as Lu_3S_4 and the intermediate sulfides of titanium (where the substructures are the hypothetical filled structures of Fig. IX.6).

A second type of ordering is total ordering such as is typified by $Ti_{5/6}O_{5/6}$, Sc_2S_3 (for the given stoichiometry the substructure (NaCl-type) is not known), Cr_5S_6 and Cr_7S_8. In these cases the vacancy ordering is complete in the sense that all of the structures have space group symmetry and all occupied sites are in principle 100% occupied. It sometimes happens, as in $Cr_{1-x}S$, that a hierarchy of ordering processes such as

$$\text{NiAs-type} \rightarrow \text{CdI}_2\text{-type} \rightarrow Cr_5S_6 + Cr_7S_8$$

will occur with decreasing temperature and some of the processes may be second-order, i.e., may occur continuously without a two-phase coexistence.

A third type of order is to an incommensurate structure, as described for $Sc_{1-x}S$. In this case the resultant structure does not have space-group symmetry (e.g., translational symmetry in some direction is destroyed). The 4th condition of Landau as discussed by Haas [42] seems to imply that strictly speaking this is the case for orderings corresponding to **k** vectors with groups of the **k** vector which allow vector invariants. It is therefore not surprising that in recent years detailed examination of diffraction patterns has frequently provided evidence of incommensurate structure (e.g., charge density waves).

With regard to the ordering phenomena discussed above two points require emphasis. One is that it is difficult to differentiate between experimental results based upon systems in the process (or quenched in the process) of ordering and those based on equilibrium systems. The second is that it is difficult to differentiate between diffraction effects from a true ordered (partial or total, commensurate or incommensurate) single crystal and those from a crystal with several domains, perhaps related by symmetry. For example, a vacancy ordering might correspond to one **k** vector in a star or to more than one in a star and the diffraction effects may average over more than one domain in a coherent substructure and produce a pattern characteristic of the latter from a system for which the ordering is characteristic of the former. For these two reasons there are structural ambiguities which are difficult, and perhaps even impossible, to resolve.

A fourth type of ordering that is especially important in octahedral oxides, but is not restricted to these cases, is that of crystallographic shear. This shear can result in a long-range ordered structure of fixed stoichiometry (such as Mo_8O_{23}) or in shear planes with a variety of spacings which to X-rays might appear to be a solid solution and on the electron microscopic level to destroy translational periodicity in one or more directions. It is interesting to note that many oxide systems have provided examples of CS, whereas sulfide systems tend to form ordered defects on a coherent substructure.

The intermediate sulfides of titanium are perhaps best described as a fifth type of vacancy ordering, however. It is possible to characterize the ordering in terms of population waves. It should, however, be noted that the stacking sequences also vary with stoichiometry, and that the two effects are apparently interrelated. As was pointed out in the original paper (×63) there is a correlation between the population waves and the stacking sequence, i.e., there appears a tendency for metal atoms to accumulate in the regions of cubic packing of sulfur atoms and for vacancies to accumulate in regions of hexagonal type packing.

In the above cases the assumption was implicitly made that if there were a displacive (atomic movement) component to the ordering, as in the CS and Ti_mS_n cases, that the movements were discrete and orderly to well defined (at least in two dimensions) positions. However, as has been recently discussed [39], there is yet another type of ordering transition which couples displacive and ordering effects in a single incommensurate transition. The idea is that in some crystals an ordering is possible only if some of the atoms in the unit cell move from their positions in the disordered structure, and the ordering results in the destruction of two sets of symmetry elements corresponding to two different irreducible representations at **k**.

The transition cannot occur continuously at **k**, according to Landau's conditions, since it would correspond to two different irr. reps. If, however, some subgroup of the symmetry elements has two partial irr. reps. at **k** and an irr. rep at $\mathbf{k} + \delta\mathbf{k}$ that are the same, then, the two different irr. rep.s at **k** are said to be "compatible" with the one at $\mathbf{k} + \delta\mathbf{k}$. It follows that the distortion can proceed to $\mathbf{k} + \delta\mathbf{k}$ in a continuous fashion, and the consequence (because of the $\delta\mathbf{k}$) is an incommensurate structure. The incommensurate nature of the transition does not result from ordering effects alone. Because of the coupling of displacive effects and order the structure has an incommensurate displacive distortion that accompanies the ordering.

IX.11 Problems

1. Provide a qualitative rationale for the observation that CsCl-type to bcc-type is observed for CnZn but not for CsCl.
2. Provide a description of an alternative vacancy ordering to $CdCl_2$-type in defect

NaCl-type that has the same cell size as $CdCl_2$-type. Describe the X-ray diffraction pattern for this alternative.

3. Discuss the comment, "Since there is no symmetry basis for the existence of the Sc_2S_3 structure to exist it is probable that the true structure has some incommensurate character" in terms of Landau's fourth condition.
4. Describe the 001, 002 and 100 X-ray diffraction pattern for Cr_5S_6.
5. Propose a structure for W_8O_{23}.

Chapter X

The Electronic Structure of Crystalline Solids

X.1 Introduction

Many important aspects of the electronic structure of crystalline solids can be understood by the application of symmetry concepts to the consideration of single electron wave functions. This chapter provides an elementary introduction to the principles and results of band theory which makes extensive use of the symmetry and reciprocal space concepts developed throughout the book. A major purpose of this chapter is to provide a rigorous but elementary overview which will enable those untutored in solid-state physics to comprehend and use the results of band structure calculations.

X.2 Band Theory

The well-known power of symmetry in guiding solutions to physical problems is incorporated into band-structure claculations at the outset. The single electron wave functions, ψ, which are determined by band theory have the property that $|\psi|^2$ has the symmetry of the lattice, that is the probability of finding a given electron in any of the volume elements related by symmetry operations is the same. However, the wave function itself can differ in sign, or in fact by multiplication by a complex quantity with unit magnitude, from point to symmetry related point, and this condition will still be met. This result is assured if ψ is a Bloch-function, that is a product of a function, φ, which has the symmetry of the solid (i.e., any symmetry operation of the solid leaves φ unchanged), and a factor $\exp(i\mathbf{k} \cdot \mathbf{r})$ where $\mathbf{r}$ is a vector in the space of the crystal and $\mathbf{k}$ is a vector such that $\mathbf{k} \cdot \mathbf{r}$ is dimensionless. That is, we let

$$\psi = \exp(i\mathbf{k} \cdot \mathbf{r})\, \varphi(\mathbf{r}) \tag{X.1}$$

where if $\mathbf{r}$ is carried into $\beta\mathbf{r} + \mathbf{t}$ by any symmetry operation of the crystal, then

$$\varphi(\mathbf{r}) = \varphi(\beta\mathbf{r} + \mathbf{t})\,. \tag{X.2}$$

It follows that electronic wave functions with different $\mathbf{k}$ vectors have different periods. Furthermore, all possible periods will be considered by the consideration of all points in a volume of reciprocal space such that $\mathbf{k} \cdot \mathbf{r} \leqq 2\pi$, i.e., within the first Brillouin zone. Thus the translational symmetry of the structure is built into the band-theory results and in the process the wave functions are labeled by **k**, the reciprocal vector that describes the periodicity of the function. As will be dicussed in what follows, the wave functions vary quasicontinuously with **k**.

The reciprocal space of a given structure has the rotational symmetry of the crystal class of that structure, and certain **k** vectors are carried into themselves under some of the rotational symmetry operations. As a trivial example, the $\mathbf{k} = 0$ vector (called the Γ point) is carried into itself under all of the rotational symmetry operations of reciprocal space. In the case of a cubic crystal in the O_h crystal class this means that the rotational symmetries appropriate to the consideration of $\mathbf{k} = 0$ wave functions are those of O_h, and it follows, for example, that the labels t_{2g} and e_g are meaningful.

A plot of the results of band-structure calculation for a cubic solid (ZrS with the NaCl-type structure) is shown in Fig. X.1 [49]. This diagram shows the energies for sets of solutions to the Schrödinger equation as functions of position along particular directions in reciprocal space (i.e., as a function of the periodicity of ψ as given by $\mathbf{k} \cdot \mathbf{r}$). Consider the point labeled $\Gamma(\mathbf{k} = 0)$. The lowest lying energy at about −0.6 Ry (1 Ry = 13.6 eV = 1,312 kJ mol^{-1}) corresponds to a wave function the angular part of which, when expanded in spherical harmonics, is made up of pure $l = 0$ (s-type) states principally centered on sulfur and is nondegenerate. The three states at 0.35 Ry are $l = 1$ (p-type) states principally centered on sulfur and correspond to the triply degenerate P_x, P_y and P_z states of O_h point group symmetry, and similarly the states of 0.52 Ry and 0.60 Ry correspond to the triply degenerate t_{2g}- and the doubly degenerate e_g-type states, respectively.

If the calculation were restricted to point Γ, where according to the above

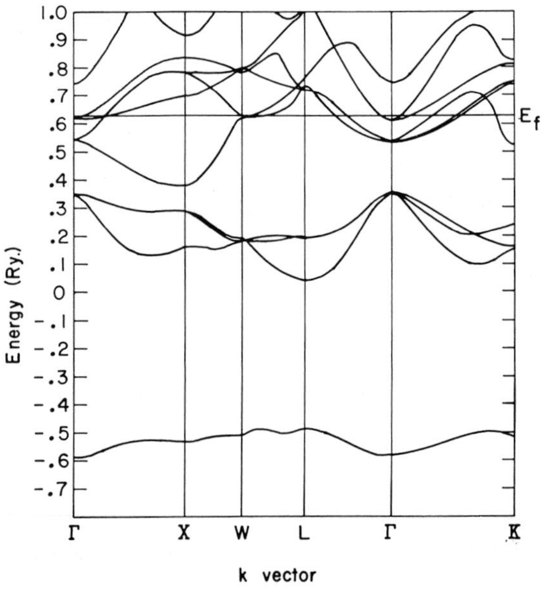

Fig. X.1. The energy bands of ZrS in the NaCl-type structure as calculated by the LAPW method.

discussion all crystal wave functions would have the translational symmetry of the lattice, the consideration of wave functions with other periodicities would not be possible. This type of calculation would be close to the molecular orbital-type of model in that the consequences of differing rotational symmetries for different translational periodicities would be ignored. In the case of band-structure calculations solutions are obtained for a large number of **k** vectors and these results provide E as a function of **k** for the different solutions to the Schrödinger equation.

The labeling along the horizontal axis of Fig. X.1 (X, W, L, Γ) identifies points of special symmetry in reciprocal space. Note that the degeneracies of different energy levels are altered at different **k** points. For example, along the line from Γ to X, which corresponds to **k** vectors lying along one of the four-fold axes, the level that is three-fold degenerate at Γ and $E = 0.35$ Ry splits into a doubly and a singly degenerate level. If that direction is taken to be the x-direction, then P_y and P_z remain degenerate (since 4-fold rotation takes P_y into P_z and vice versa) but P_x is symmetrically inequivalent to P_y and P_z in the absence of the 3-fold rotational operations that are appropriate at point Γ. It is also important in understanding Fig. X.1 to realize that for $\mathbf{k} \neq 0$ along Γ to X (and elsewhere in reciprocal space) mixing occurs, e.g., both P_x and P_z and d_{xy} and d_{xz} represent doubly degenerate functions corresponding to the same symmetry (same irr. rep.) and thus the valence band can, by symmetry, take on some d-charakter and the "d-band" can take on some p-character. This mixing is a general characteristic of much of reciprocal space. Figure X.1 does not portray E vs **k** for all of reciprocal space, but only a sampling along the directions connecting the points of high symmetry.

The E vs **k** results can be analyzed in a variety of ways. One common analysis is to determine, by computation, the number of states (i.e., the number of discrete solutions to the Schrödinger equation in the quasicontinuum of solutions) in an energy interval between E and $E + \delta E$, and when this number is divided by δE the density

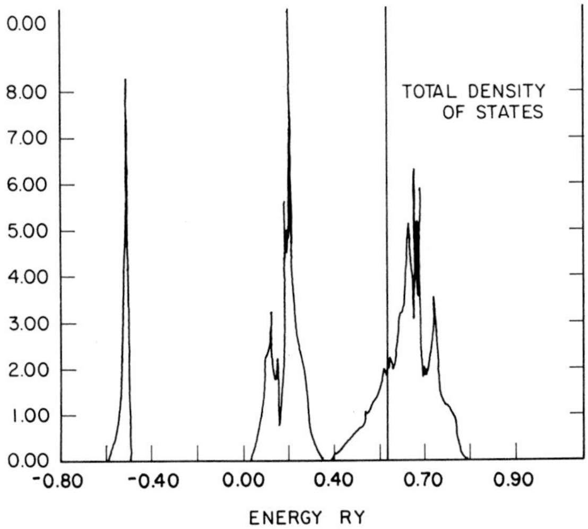

Fig. X.2. The density of states of ZrS in the NaCl-type structure.

of states (d.o.s) at E has been determined. A plot of the d.o.s., dN/dE, vs E is shown for ZrS in Fig. X.2. High densities of states correspond to relatively slow variations of E with **k**, i.e., rather flat curves in Fig. X.1.

Since the functions with different rotational symmetry at point Γ can mix at other **k** points it is interesting to inquire about the symmetry (e.g., in terms of spherical harmonics, the l-character) of the various states that constitute the density-of-states curve. This has been done for ZrS, i.e., the ψ functions have been expressed as combination of spherical harmonics centered on Zr and S sites, and the s, p, d and f character of the ψ functions determined. It was then possible to determine angular decomposed densities of states as shown in Fig. X.3. This analysis provides the following interpretation:

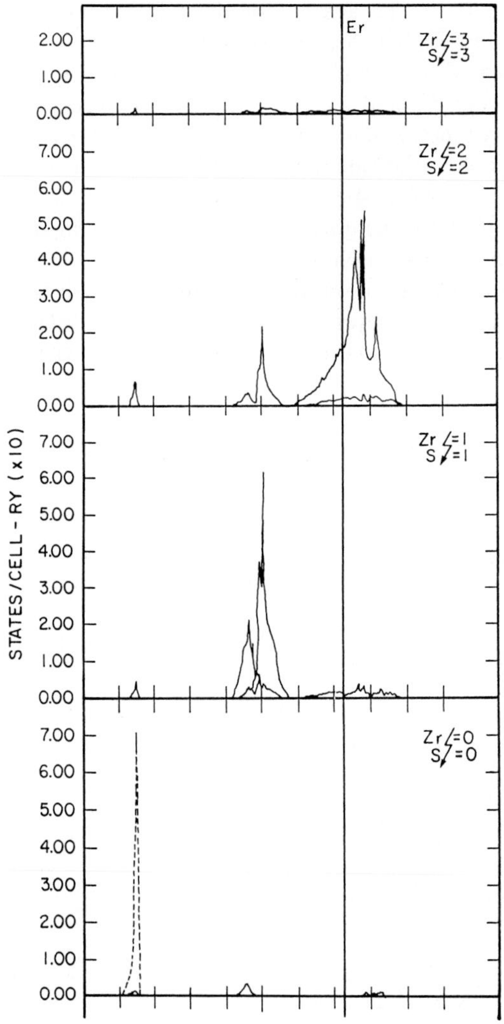

Fig. X.3. The l-decomposed densities of states for ZrS in the NaCl-type structure

1. the lowest lying band (a relatively flat band) centered at —0.52 Ry is made up principally of sulfur s-type (l = 0) states with very small contributions of zirconium l = 0, 1, 2 and 3-type states.

2. the valence band centered at 0.20 Ry is principally sulfur p-type (l = 1) states with a rather substantial contribution from zirconium d-type states, and small contributions from zirconium s, p and f-type states, and

3. the d-band between 0.35 Ry and 0.62 Ry (the highest filled states corresponding to the Fermi energy of 0.62 Ry) is principally zirconium d-type with small contributions from zirconium p- and sulfur d-type states.

It is instructive to compare this analysis with that given for the M.O. scheme of Fig. I.2, and to note that so far as qualitative discussion is concerned, the

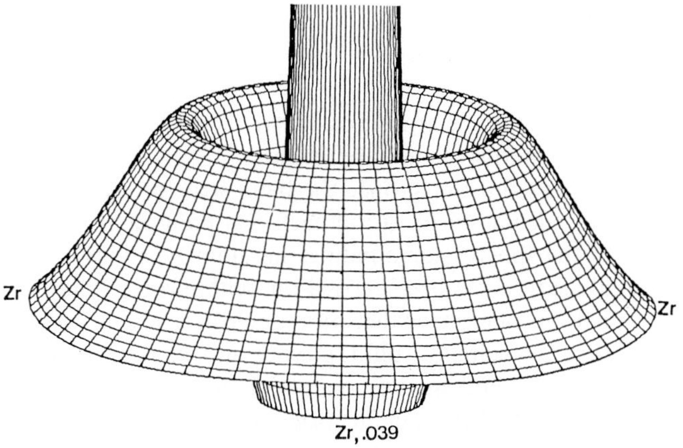

Fig. X.4. The calculated electron density distribution for the sulfur sphere in ZrS with the NaCl-type structure.

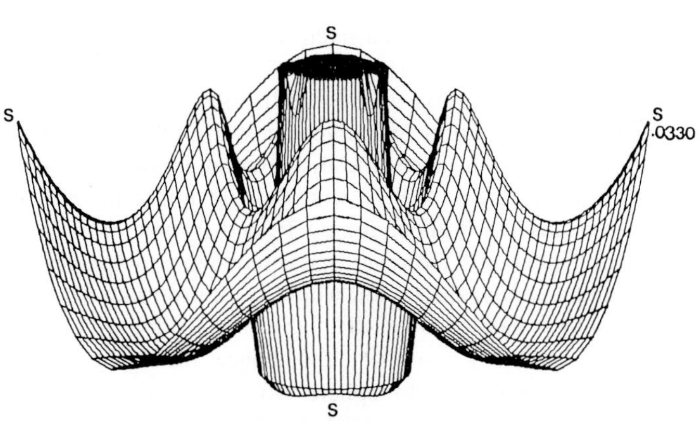

Fig. X.5. The calculated electron density in the Zr sphere in ZrS with the NaCl-type structure.

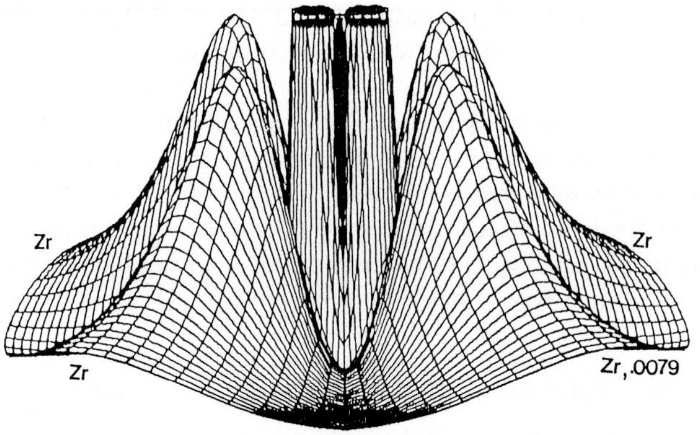

Fig. X.6. The calculated electron density distribution for the d-band states only in the Zr sphere in ZrS with the NaCl-type structure.

discussion of interactions in rock-salt type transition-metal monochalcides is not greatly different for the M.O. and band-structure approaches.

The availability of electronic wave functions, once the band-structure calculation has been completed, enables the calculation of $|\psi|^2$ at each real point, and thus the electron density distribution in the spheres surrounding the atom positions can be obtained. Furthermore, this calculation has been carried out within energy intervals so that, for example, the distributions of electron densities for electrons in the valence band region in the sulfur sphere (principally sulfur p-type) and in the zirconium sphere (principally zirconium d-type) and in the conduction band region for the zirconium sphere (principally zirconium d-type, in fact principally t_{2g}-type) have been calculated and plotted (Figs. X.4, 5 and 6, respectively). These results demonstrate the extent of nonsphericity of the electron density distribution, and these non-sphericities provide a possible meaning to the directed chemical bond concept appropriate to the solid state.

It seems reasonable to suggest that if the electron density distribution is spherically symmetric within a sphere of reasonable size about a given atom position, as is the case for S in ZrS (and in ScS [50], TiS and VS [51] as well) then the electron density distribution makes no directional contribution to the stabilization of that structure. On the other hand, the aspheric distribution in the sphere surrounding zirconium is, it is suggested, an indication that the elctronic distribution in the zirconium sphere has an effect upon the symmetry of ZrS. A first thought might be that the lifting of the degeneracy such as discussed above for P_x, P_y and P_z-type orbitals for $\mathbf{k} \neq 0$ would play a role in the asphericity. However that discussion was appropriate to a given (the x) direction, and a similar discussion would be appropriate for the y and z directions. When the sum is carried over all of the first Brillouin zone the results is equal participation by P_x, P_y and P_z and hence the (very nearly) spherical electron density distribution in the sulfur sphere. Any slight asphericity must result from the very small d- and f-type contributions to the electronic wave functions in the sulfur spheres.

However, since the predominant contributions in the zirconium spheres have

$l = 2$ character, and since these $l = 2$ functions are not 5-fold degenerate, but are split into 3-fold degenerate t_{2g}-type and 2-fold degenerate e_g-type functions by the cubic symmetry, it follows that the electron density distribution in the zirconium sphere will not be spherically symmetric unless there is by chance an equal filling of all five d-type orbitals. It thus appears that the t_{2g}-e_g splitting is the most important effect in leading to an aspherical electron density distribution and hence to the directional bonding effects in ZrS.

It should be remarked that CaS with no important d-electron contributions also crystallizes in the rock-salt type structure, and thus it is clear that $l = 2$ electronic functions and the resulting t_{2g}-e_g splitting and consequent electron density asphericity are not necessary for the formation of this structure type, but rather that these features are consistent with the rock-salt structure and important to the transition-metal containing compounds. A rough estimate of the effect of the d-electrons upon the rock-salt structure can be obtained by comparing the atomization enthalpies of CaS, ScS and TiS which are 933 kJ [52], 1071 kJ [53] and 1025 kJ [54], respectively. It thus appears that the d-electrons are, at least for the early transition metal sulfides, weakly bonding. For CaO and TiO the atomization enthalpies are 1059 kJ and 1239 kJ [55], respectively, suggesting a somewhat stronger d-electron bonding effect in TiO.

It is suggested that three effects of adding d-electrons to the rock-salt structure can be discerned:

1. an increase in the directional nature of the metal contribution to the M—X interaction,
2. an increase in the M—M interactions, and
3. an increase in the interaction between the t_{2g} electrons and the M—X bonding electrons.

The first two effects, it would be guessed, tend to stabilize the structure and the third would have a destabilizing effect. The significance of the third effect is inferred from the fact that the conduction band electron density distribution has its maximum directed through the centers of the triangular faces of the octahedra of the surrounding sulfur atoms (and this was also found to be true in the cases of TiS and VS [51]), as becomes apparent from an analysis of figures such as Fig. X.6. It thus results that the traditional view of the stabilization of octahedral coordination of transition metals in terms of the ligand-field model (covalent e_g-ligand and repulsive t_{2g}-ligand interactions) is substantiated by the detailed analysis made possible by the band-theory results. The t_{2g}-type orbitals are located at the Fermi level and appear in the electron density plots as the states which provide for delocalization of the electrons over the metal-atom sites providing both the energetic and the geometric rationale for metallic conduction in ZrS (and related rock-salt type compounds).

The results of band theory can help to provide answers to the essential chemical question of what electronic effects stabilize one symmetry (structure) over another. Currently this can be rigorously accomplished in practice only for the elemental metals because of the computer time required to carry out the self-consistent calculation of the total energies. Important qualitative features can be discerned without such extensive calculations, however.

The case of ZrS is a good example. The current experimental view of the monosulfides of zirconium is that while stoichiometric ZrS in the rock-salt structure can perhaps be obtained by rapid quenching, the stable form of ZrS in the rock-salt structure is obtained when $\sim 20\%$ of the cation sites are vacant, and then the vacancies order on the rock-salt lattice. Thus the calculation described above is for metastable ZrS, whereas stable ZrS (or, as it is believed, slightly anion deficient ZrS, i.e., $ZrS_{0.99}$) crystallizes in the WC structure (Fig. IV.3). Since the qualitative results of the band-structure calculation for the rock-salt structure case of ZrS are in excellent agreement with the general M.O. and ligand-field models for group IV transition-metal chalcides, it would seem to be worthwhile to examine these models for the WC case.

However, when we seek examples of such modeling or attempt to construct such models we find that it is not at all clear how to proceed, and we find by this counter example that the success of the simple models for the rock-salt structure is highly dependent upon the high symmetry of this structure, which provides very few alternatives for first approximation to the electronic configurations of the atoms.

Thus, we have turned to band theory and the results for WC-type ZrS [49] are shown in Fig. X.7. The resulting d.o.s. plot (Fig. X.8) can be compared with that for the rock-salt case (Fig. X.2) and it is apparent that there are similarities (a low-lying s-band, an intermediate s valence band and a higher lying d-band). Fortunately there are also some important differences, or else it would be clear that these results are not helpful in the consideration of the structure.

The most readily apparent difference in the total d.o.s. (Figs. 2 and 8) is in the d-band region which is sharply peaked just below E_F in the WC-case and which gradually rises through E_F in the NaCl-case. The partial d.o.s. curves [49] show

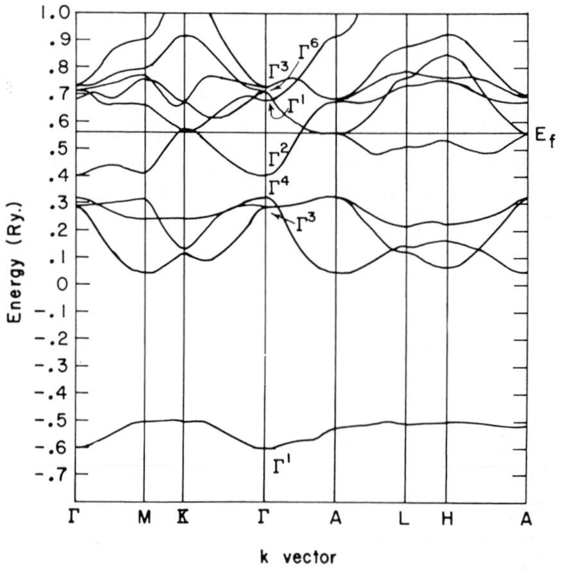

Fig. X.7. Energy bands for ZrS with the WC-type structure.

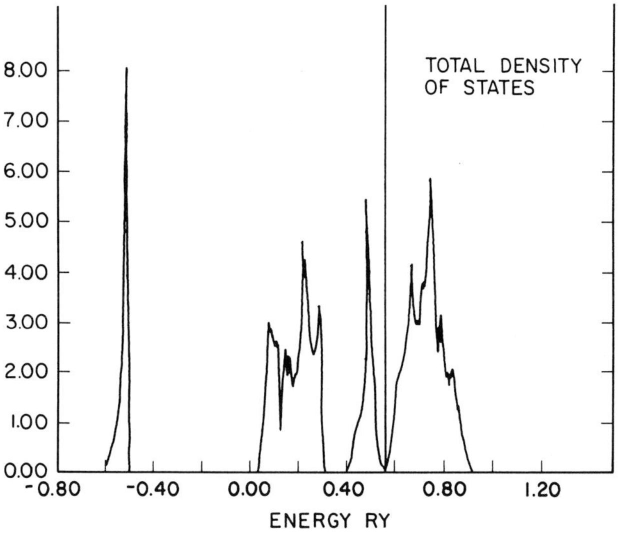

Fig. X.8. The d.o.s. for ZrS with the WC-type structure.

that these states are in fact principally Zr l = 2 type with minor contributions from S l = 2, 1, 0 and Zr l = 1 states. The electron density distribution plots for the WC-case are shown in Figs. X.9 and 10. Figure X.10 shows that the conduction band states, i.e., the states in the interval 0.40 Ry $\leqq E \leqq E_F$, give rise to an electron density distribution that is peaked in the a-b plane in the direction through the

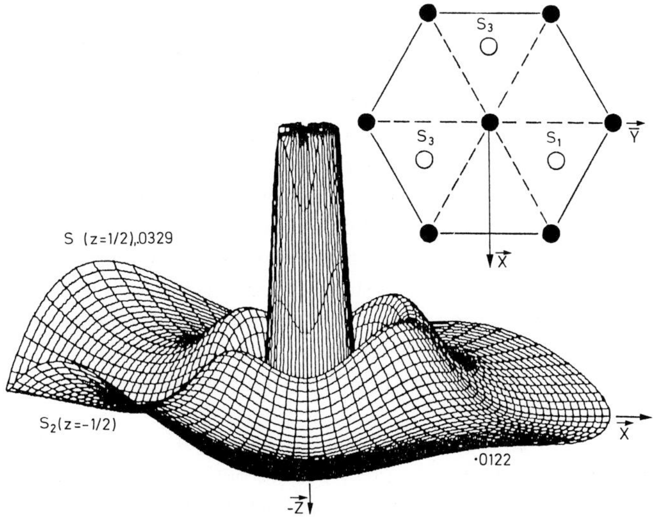

Fig. X.9. The calculated electron density distribution for p-band states in the zirconium sphere in ZrS with the WC-type structure.

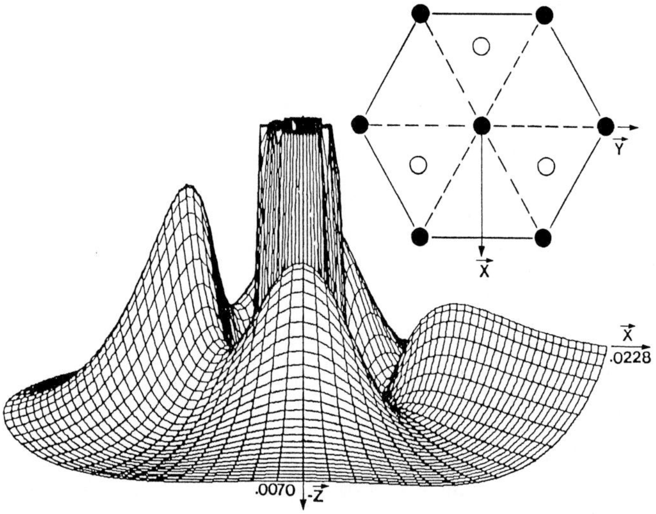

Fig. X.10. Calculated electron density distribution for d-band states of ZrS with the WC-type structure (Zr sphere).

centers of the rectangular faces of the trigonal prism composed of sulfur atoms surrounding the central atom. Thus in the WC-type structure the peak in the (principally) Zr d d.o.s. is associated with Zr—Zr interactions in delocalized 3-center bonding networks in which the electron density maxima are in directions which maximally avoid the ligands.

Taken together the NaCl-type and WC-type results suggest an interpretation for the d-orbital interactions in these solids. In both cases there are d-orbital interactions which result in metallic properties and, presumably, tend to stabilize the structures (although to what extent is unknown). In the case of the NaCl-type structure the interactions are to 12 nearest Zr neighbors with a directional preference shown for interaction through the 8 faces of the coordination octahedron. In the case of the WC-type structure the principal directionality is to 6 nearest Zr neighbors in the a-b plane (and not to the two along c), again through the faces of the coordination polyhedron, in this case a trigonal prism. It is reasonable to suggest that the conduction electrons owe an important part of their asphericity to repulsive interactions with the sulfur atoms (and/or the Zr-S bonding electrons). It follows that one term to be considered in thinking about structure-stability relationships is the relative destabilization of the structures by the local repulsive interactions of valence-band and d-band electrons, and hence the relative flexabilities of the d-band states in conforming to the symmetries of the coordinating atoms in different types of structure emerges as an important consideration.

An examination of the valence-band electron densities indicates that the covalent Zr—S interactions are similar in the two structure types. In both cases the electron densities in the sulfur spheres out to 144 pm are spherically symmetric to within a high degree of precision. It can be seen that there are in both structure types important Zr—S interactions to which the principal directional contribution come from the

Zr d-type states. The significance of the differences between the valence band interactions is not easy to discern.

The results for TiS and VS [51] in the NiAs-type structure (Fig. IV.4), are consistent with those given above. The metal-metal interactions are delocalized with the electron density maxima appearing at directions through the centers of the faces of the coordinating trigonal antiprisms (trigonally distorted octahedra). A comparison of the electron density distribution in VS and in TiS indicates a decrease in asphericity with increasing number of d-electrons. Perhaps part of the significance of the return to the NaCl-type structure with MnS is the tendency toward spherical symmetry for Mn owing to the half-filled d^5 configuration. Based upon the above, a partial analysis of the factors active in determining crystal structures which is based upon the results of band theory can be provided. It has long been recognized that while the rock-salt structure Fig. I.1 occurs frequently for compounds with negligible metal d-metal d interactions (alkali halides, alkaline-earth chalcides, rare-earth pnictides) as well as for a few 3-d metal compounds some of which (e.g., TiO) are metallic, the NiAs-type structure (Fig. IV.4) occurs predominantly for compounds for which metal d participation is important (e.g., metallic transition-metal sulfides, selenides, etc.). There is no known case of a compound which is stable in two polymorphic forms one of which is NaCl-type and the other of which is NiAs-type. The comparison of NaCl-type and NiAs-type solids provides a basis for testing our understanding of the relationship of d-band states to crystal structure.

One early interpretation of the relationship between d-electrons and stability of the NiAs-type structure relative to NaCl-type rested upon the observation that the metal atom substructure is trigonal-prismatic in the NiAs-case and is octahedral in the NaCl case. One consequence of this difference in substructure is that it is possible in the NiAs structure type for the metal atoms to come in closer contact (along the c axis) for a given metal to nonmetal distance than is the case in the NaCl-type

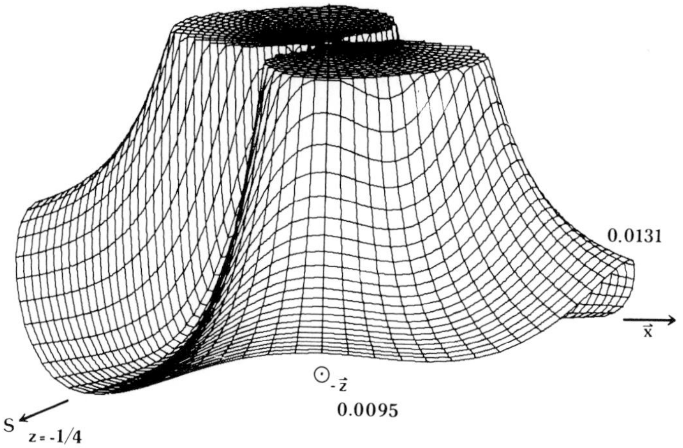

Fig. X.11. Calculated electron density distribution in the Ti sphere of TiS (NiAs-type), d-band states only.

structure. For example the short Ni-Ni distance along the c axis in high temperature NiS with the NiAs-type structure is 267.5 pm [19] whereas the Ni—Ni distance in hypothetical NiS with the NaCl-type structure with the same Ni—S distance is 339 pm. The suggestion based upon this observation is that d-electron involvement favors the NiAs-type structure trough d_{z^2} interactions along the c-axis while significant ionic character destabilizes the NiAs-type structure relative to the NaCl type because of the cation-cation repulsions (the Madelung constant for the NaCl-type structure with the same metal-nonmetal distance is always greater then that for the NiAs-type structure).

The compound TiS with the NiAs type structure provides an excellent test case for considering the necessity of d_{z^2} bonding in stabilizing the NiAs structure type. ScS forms the NaCl-type structure and TiS forms the NiAs-type structure and thus it appears that the additional d-electron in TiS has tipped the balance in favour of the latter structure type. However, c/a for TiS is 1.93, to be compared with the values for the vast majority of the over 40 NiAs-type binary solids listed in Pearson [26] for which c/a is less than the ideal, hcp anion, value of $\sqrt{\frac{8}{3}} = 1.633$.

From the point of view developed in this chapter a test of the relationship of d-electrons to structure is provided by the results of a band structure calculation. In particular the d-band electron density plot shown in Fig. X.11 shows that the z direction (directly out of the page) is not the direction of maximum electronic interaction. As mentioned previously the maxima in d-d electronic interactions are in the directions through the faces of the sulfur coordination polyhedron (a trigonal antiprism) other than those perpendicular to c. Thus, a significant d_{z^2} interaction is seen not to be necessary for the formation of the NiAs structure. What appears to be the case is that the symmetry change (from NaCl-type to NiAs-type) permits a relative localization of the d-electron density distribution (six and two directions rather than eight all equivalent) and this localization, together with the variable c/a ratio in the NiAs-type structure, provides for a greater separation of d-band and valence band electrons without eliminating the d-electron interactions. Working through the geometry for TiS (a = 330 pm, c = 638 pm [19]) yields, assuming the conduction electron density maximum at the center of the polyhedral face to be equidistant from all three anions, a distance of 207.4 pm in the NiAs case and 202.8 pm for the case of the NaCl structure with the same metal-nonmetal distance (248.0 pm). Furthermore this distance in the NiAs structure increases with c/a, i.e., the large c/a for TiS is favored by the repulsion between d-band and valence band states.

It thus appears that the d-electron interactions involved in stabilizing the NiAs-type structure are not necessarily d_{z^2} interactions, but include those of the other d orbitals, including contributions from the repulsive interactions.

X.3 The Nearly Free Electron Model of Band Structure

Some aspects of the behavior of electrons in crystalline solids can be understood in terms of a detailed treatment on the very elementary level of the nearly free electron model. Much of this understanding is in terms of the theory of space groups, which has been developed in detail in Chapter VI, and the interaction of this group theory with quantum mechanics, which is the subject of this chapter. For a good understanding of the material of this chapter it is important to have become familiar with the material of Chap. VI.

Except in the case of accidental degeneracy, the eigenfunctions of the single electron Hamiltonian

$$\mathscr{H} = -\frac{\hbar^2}{2m}\left(\frac{\partial^2}{\partial x^2} + \frac{\partial^2}{\partial y^2} + \frac{\partial^2}{\partial z^2}\right) + V(\mathbf{r}) \tag{X.3}$$

corresponding to a given eigenvalue are basis functions for a single irreducible representation of the group of operations that leave $\mathscr{H}$ invariant. The proof of this statement follows. The potential of a solid presumably has the symmetry of the crystalline solid (of the space group), i.e., if $\beta|t$ is a symmetry operation of the space group (g) then $V(\mathbf{r}) = V(\beta\mathbf{r} + \mathbf{t})$. It follows by inspection that $\mathscr{H}$ is invariant under operations of the space group. If $\mathscr{P}$ is an operator which transforms any function according to one of the symmetry operations of g then, if

$$\mathscr{H}\psi_n = E_n\psi_n \tag{X.4}$$

(i.e., if ψ_n is an eigenfunction with the eigenvalue E_n), it follows from the fact that $\mathscr{H}$ is invariant under $\mathscr{P}$ that $\mathscr{P}$ and $\mathscr{H}$ commute

$$\mathscr{P}(\mathscr{H}\psi_n) = \mathscr{H}(\mathscr{P}\psi_n)\,. \tag{X.5}$$

Thus

$$\mathscr{H}(\mathscr{P}\psi_n) = \mathscr{P}(\mathscr{H}\psi_n) = \mathscr{P}E_n\psi_n = E_n\mathscr{P}\psi_n\,. \tag{X.6}$$

This means that if ψ_n is an eigenfunction of $\mathscr{H}$ with eigenvalue E_n and $\mathscr{P}$ is an operator which transforms ψ_n according to $\beta|\mathbf{t}$, one of the symmetry operations of g, then $\mathscr{P}\psi_n$ is also an eigenfunction of $\mathscr{H}$ with eigenvalue E_n, i.e., ψ_n and $\mathscr{P}\psi_n$ are either the same function within a multiplicative constant or are degenerate functions.

If the eigenvalue is l-fold degenerate then there are l functions ψ_{n1}, ψ_{n2}, $\psi_{n3}, \ldots, \psi_{nl}$ given by ψ_n and its transformations, and a solution of the Schrödinger equation can be expressed as a linear combination of these functions. Therefore we can write in general for $\mathscr{P}_1\psi_{nk}$

$$\mathscr{P}_1\psi_{nk} = \sum_{j=1}^{l} a_{jk}\psi_{nj} \tag{X.7}$$

and

$$\mathscr{P}_2\psi_{nj} = \sum_{m=1}^{l} b_{mj}\psi_{nm} \tag{X.8}$$

and, applying $\mathscr{P}_2$ to $\mathscr{P}_1\psi_{nk}$,

$$\mathscr{P}_2\mathscr{P}_1\psi_{nk} = \sum_{j=1}^{l} a_{jk} \sum_{m=1}^{l} b_{mj}\psi_{nm} \tag{X.9}$$

$$= \sum_{j=1}^{l} \sum_{m=1}^{l} a_{jk}b_{mj}\psi_{nm} . \tag{X.10}$$

Now if $\mathscr{P}_2\mathscr{P}_1 = \mathscr{P}_3$ then

$$\mathscr{P}_3\psi_{nk} = \sum_{m=1} c_{mk}\psi_{nm} \tag{X.11}$$

and comparison yields

$$c_{mk} = \sum_{j=1} b_{mj}a_{jk} . \tag{X.12}$$

This is exactly the way matrices containing elements a_{jk} and b_{mj} combine to form the c_{mk} matrix and thus it follows that these matrices, for which the ψ_{ni} are the basis functions, form a representation of the group of operations (the space group) that leaves $\mathscr{H}$ invariant. That these matrices are irreducible follows from the fact that it would otherwise be possible to form sets of linear combinations of the eigenfunctions which reduced the representation, but the corresponding eigenvalues would then have no reason of symmetry to be the same and would therefore (unless by accident) differ, which would contadict the original assumption that we were dealing with an energy level that was l-fold degenerate. This completes the proof.

Now it is possible to consider as the space group in question the subgroup of pure translations $\{\varepsilon|T_i\}$. This subgroup contains operators all of which commute (pure translational symmetry operations combine irrespective of order). In such a group the number of classes, and therefore the number of irr. rep.s, equals the number of elements, and the requirement that the sum of the squares of dimension of the irr. rep.s equals the number of group elements then yields the result that such representation must be one-dimensional.

Combination of the results just obtained leads to the result that the eigenfunctions of the translations are singly degenerate, and thus transform into themselves multiplied by a constant under the pure translational symmetry operations. This fact together with the boundary conditions placed upon the single electron wave functions, which will now be discussed, serve to determine in part the nature of these wave functions.

The number of single electron states associated with a crystal of finite size is determined by the boundary conditions placed upon the single electron wave functions. Because the bulk properties of macroscopic crystalline solids are independent of the size and shape of the crystal, the calculated electronic states are independent of the exact nature of the boundary conditions, i.e., an arbitrary means of limiting the number of states through the imposition of boundary conditions will lead to the correct results for the bulk solid.

The scheme that has been adopted is to suppose that

$$\psi(\mathbf{r}) = \psi(\mathbf{r} + N_1\mathbf{a}) = \psi(\mathbf{r} + N_2\mathbf{b}) = \psi(\mathbf{r} + N_3\mathbf{c}) \qquad \text{(X.13)}$$

where N_1, N_2, N_3 are very large integers. This supposition amounts to assuming that the properties of crystals made up of very large parallelepiped blocks ($N_1\mathbf{a} \times N_2\mathbf{b} \times N_3\mathbf{c}$) in which the wave functions are identical are approximations to the properties of real crystals. This assumption, which is intuitively reasonable, can be checked by varying the size of N_1, N_2 and N_3, and is known with certainty to introduce no extraneous effects large ($\sim 10^{23}$) values of N_i.

These Born-von Karman boundary conditions coupled with the fact that ψ is a basis function for a 1 — D irr. rep. of the translations means that translation by a translational symmetry operation, e.g., **a**, takes ψ into $C_1\psi$, and

$$C_1^{N_1} = 1\,, \qquad \text{(X.14)}$$

or

$$C_1 = \exp(2\pi i p_1/N_1)\,, \qquad \text{(X.15)}$$

where p_1 is an integer between 0 and N — 1 (such that $C_1^{N_1} = \exp(2\pi i p_1) = 1$). Similar results hold for **b** and **c**, and thus $\mathbf{T} = n_1\mathbf{a} + n_2\mathbf{b} + n_3\mathbf{c}$ takes C_i into $C_i^{n_i}$ and thus ψ into $\exp[2\pi i\{(p_1n_1/N_1) + (p_2n_2/N_2) + (p_3n_3/N_3)\}]$ times ψ. In this expression N_1, N_2 and N_3 are very large and fixed by the size of the parallelepiped domain chosen for the crystal, p_1, p_2 and p_3 are triples of integers each one of which may be specified between O and N_i — 1, and the specification of the triples specifies a given irr. rep. of the translations, and n_1, n_2 and n_3 correspond to a chosen pure translational symmetry operation.

If a wave vector **k** is defined by

$$\mathbf{k} = \left(\frac{p_1}{N_1}\right)\mathbf{a}^* + \left(\frac{p_2}{N_2}\right)\mathbf{b}^* + \left(\frac{p_3}{N_3}\right)\mathbf{c}^* \qquad \text{(X.16)}$$

then for $\mathbf{T} = n_1\mathbf{a} + n_2\mathbf{b} + n_3\mathbf{c}$,

$$\frac{p_1n_1}{N_1} + \frac{p_2n_2}{N_2} + \frac{p_3n_3}{N_3} = \mathbf{k}\cdot\mathbf{T}\,, \qquad \text{(X.17)}$$

and **T** is seen to take $\psi_\mathbf{k}$ into $(\exp 2\pi i\mathbf{k}\cdot\mathbf{T})\,\psi_\mathbf{k}$. Note that the cyclic boundary conditions imply that only a finite (although very large) number of **k** vectors correspond to allowed representations (eigenfunctions), namely, reciprocal space is divided into $N_1 \cdot N_2 \cdot N_3$ (small) parallelepipeds defined by $\mathbf{a}^*/N_1$, $\mathbf{b}^*/N_2$ and $\mathbf{c}^*/N_3$ and thus reciprocal space is rendered quasicontinuous by the boundary conditions.

A function which is of the form such that **T** takes $\psi_\mathbf{k}$ into $(\exp 2\pi i\mathbf{k}\cdot\mathbf{T})\,\psi_\mathbf{k}$ is

$$\psi_\mathbf{k}(\mathbf{r}) = (\exp 2\pi i\mathbf{k}\cdot\mathbf{r})\,\varphi_\mathbf{k}(\mathbf{r})\,, \qquad \text{(X.18)}$$

where $\varphi_{\mathbf{k}}(\mathbf{r})$ is a function which is totally symmetric with respect to translation, i.e.,

$$\varphi_{\mathbf{k}}(\mathbf{r} + \mathbf{T}) = \varphi_{\mathbf{k}}(\mathbf{r})\,. \tag{X.19}$$

The subscript **k** labels $\psi_{\mathbf{k}}$ because the "plane wave" part, i.e., exp $2\pi i\mathbf{k} \cdot \mathbf{r}$, of the eigenfunction depends explicitly upon **k**, and because in the general case the nature of $\varphi_{\mathbf{k}}(\mathbf{r})$ depends upon the irr. rep. of the space group, i.e. upon the small representation, which in turn depends upon **k**. Functions of this form are called Bloch functions.

For some electrons within a crystalline solid the potential is of such a magnitude relative to the kinetic energy that the electrons behave as nearly free particles, slightly perturbed by the periodic potential. The model adopted for this case is to take the single electron Hamiltonian with V = 0 as describing the behavior of the electrons throughout much of the Brillouin zone, but to explicitly consider the potential as a perturbation in the neighborhood of the zone boundary. The Bloch form of the single electron wave functions is, as shown above, a requirement of the three-dimensional periodicity of the lattice:

$$\psi_{\mathbf{k}}(\mathbf{r}) = [\exp 2\pi i\mathbf{k} \cdot \mathbf{r}]\, \varphi_{\mathbf{k}}(\mathbf{r})\,, \tag{X.20}$$

and with $\mathbf{r} = x\mathbf{a} + y\mathbf{b} + z\mathbf{c}$ and $\mathbf{k} = \left(\frac{\mathscr{P}_1}{N_1}\right)\mathbf{a}^* + \left(\frac{\mathscr{P}_2}{N_2}\right)\mathbf{b}^* + \left(\frac{\mathscr{P}_3}{N_3}\right)\mathbf{c}^*$,

$$\psi_{\mathscr{P}_1, \mathscr{P}_2, \mathscr{P}_3}(x, y, z) = \left\{\exp\left[2\pi i\left(\frac{\mathscr{P}_1 x}{N_1} + \frac{\mathscr{P}_2 y}{N_2} + \frac{\mathscr{P}_3 z}{N_3}\right)\right]\right\} \phi_{\mathbf{k}}(\mathbf{r})\,. \tag{X.21}$$

In order to avoid unproductive geometrical detail it will be assumed that **a**, **b** and **c** are equal in length and mutually orthogonal (cubic reciprocal lattice) and note that, taking $\nabla^2 = \frac{1}{a^2}\left(\frac{\partial^2}{\partial x^2} + \frac{\partial^2}{\partial y^2} + \frac{\partial^2}{\partial y^2}\right)$ so that x, y, z are in units of a,

$$\begin{aligned} &-\frac{\hbar^2}{2ma^2}\left[\frac{\partial^2}{\partial x^2} + \frac{\partial^2}{\partial y^2} + \frac{\partial^2}{\partial z^2}\right] \exp 2\pi i\left(\frac{\mathscr{P}_1 x}{N_1}\,\frac{\mathscr{P}_2 y}{N_2}\,\frac{\mathscr{P}_3 z}{N_3}\right) \\ &= \frac{\hbar^2}{2ma^2}\left[\left(\frac{\mathscr{P}_1}{N_1}\right)^2 + \left(\frac{\mathscr{P}_2}{N_2}\right)^2 + \left(\frac{\mathscr{P}_3}{N_3}\right)^2\right] \cdot \\ &\times \exp\left[2\pi i\left(\frac{\mathscr{P}_1 x}{N_1} + \frac{\mathscr{P}_2 y}{N_2} + \frac{\mathscr{P}_3 z}{N_3}\right)\right] \end{aligned} \tag{X.22}$$

i.e., $\psi_{\mathbf{k}}$ with $\varphi_{\mathbf{k}}(\mathbf{r}) \equiv 1$ is an eigenfunction of both the translational symmetry operators and the kinetic energy operator with eigenvalues

$$E_{P_1, P_2, P_3} = \frac{\hbar^2}{2ma^2}\left[\left(\frac{\mathscr{P}_1}{N_1}\right)^2 + \left(\frac{\mathscr{P}_2}{N_2}\right)^2 + \left(\frac{\mathscr{P}_3}{N_3}\right)^2\right]. \tag{X.23}$$

N_1a, N_2a and N_3a are the dimensions of the macroscopic crystal unit, hence taking $N_1a = N_2a = N_3a = L$,

$$E_{p_1, p_2, p_3} = \frac{h^2}{2mL^2}(\mathscr{P}_1^2 + \mathscr{P}_2^2 + \mathscr{P}_3^2), \quad \text{(X.24)}$$

the energy levels of a particle in a cubic crystal of length L. Neglecting the perturbation at the Brillouin zone edge for the moment we can plot E vs p and, owing to the very large size of N, obtain a quasicontinuous plot (Fig. X.12).

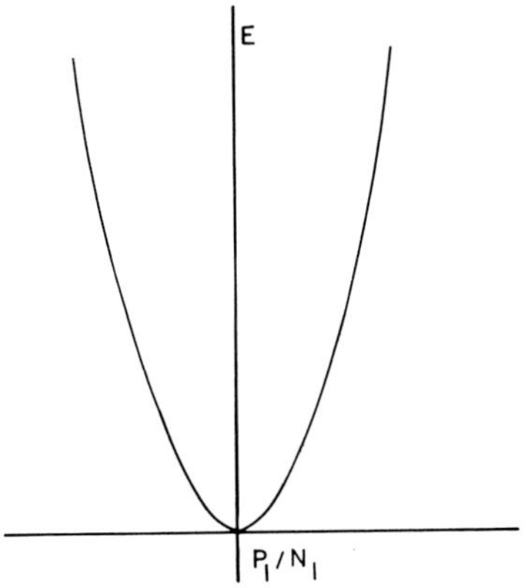

Fig. X.12. Energy vs wave vector for a free electron.

We know that p_1/N_1 is the wave vector component in the **a*** direction, and that **a*** is a reciprocal lattice vector, and thus that E vs p_1/N_1 is periodic in, for example, the **a*** direction and has the appearance shown in Fig. X.13. However, it is necessary to consider the pertubing effect of the potential at the zone boundary (at $\mathbf{k} = \mathbf{a}^*/2$). According to Fig. X.13, based upon complete neglect of V, the doubly degenerate free electron-like wave functions at $\mathbf{k} = \pm \frac{\mathbf{a}^*}{2}$ are $\exp(\pi i x)$

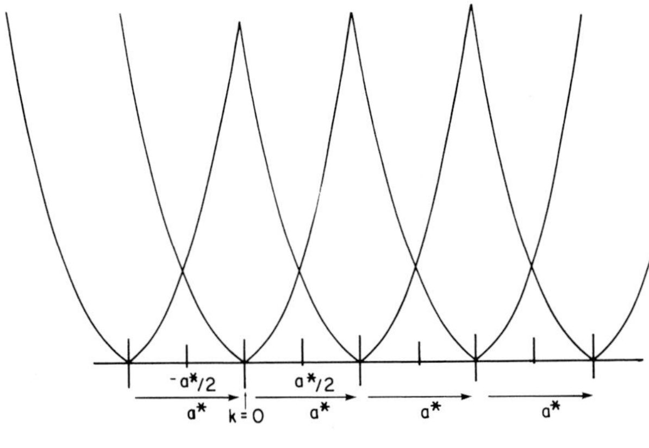

Fig. X.13. E vs k along **a***.

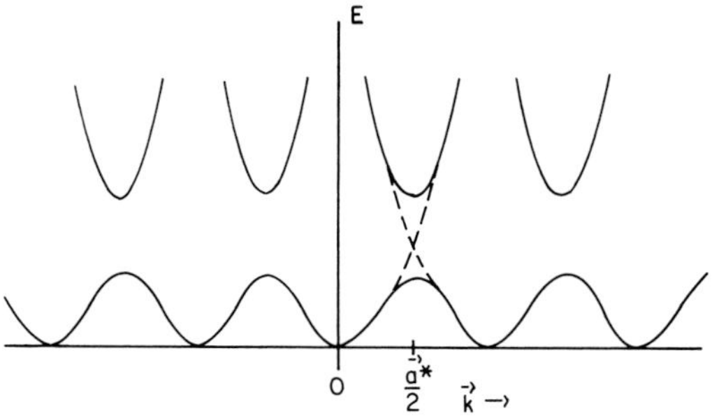

Fig. X.14. E vs. **k** along **a*** showing the effect of a periodic potential.

and exp ($-\pi ix$). However, barring the existence of a 2_1 axis along x (as discussed in Section VI.4) the two basis functions under consideration here are bases for a reducible representation at $\mathbf{k} = \mathbf{a}^*/2$, i.e., $\cos \pi x = \frac{\exp(\pi ix) + \exp(-\pi ix)}{2}$ and $\sin \pi x = \frac{\exp(\pi ix) - \exp(-\pi ix)}{2i}$ are basis functions for two one-dimensional representations Detailed analysis [56] shows that introducing the potential as a perturbution allows mixing of the two states in the neigborhood of $\mathbf{a}^*/2$, and thus an E vs. **k** curve as shown in Fig. X.14.

X.4 Conclusion

As described in the first section of this chapter in more realistic calculations attempts are made to calculate an effective potential which has the symmetry of the lattice, and solve the Schrödinger equation including the potential. The results can be expressed as band plots, such as those shown for ZrS. There are a number of features of these plots that can be understood in terms of concepts that have been provided at various places throughout the text. First is the concept of the Brillouin zone, the unit cell in reciprocal space within which **k** points corresponding to all possible irr. reps. are contained. Second is the fact that the Schrödinger equation yields eigenvalues (E_n, n = 1, 2, 3, ...) specific to **k** points within the zone. The third is the idea of semicontinuum of eigenstates in reciprocal space leading to bands of states which can be considered by plotting E_n vs. k in reciprocal space. The fourth is that the point group of the wave vector at each **k** point determines the nature of the irr. reps., and hence the degeneracies of eigenstates at **k** points, and also the compatibilities of states at special **k** points and those neighboring them. Finally, the nearly free electron behavior in some bands can be seen in nearly parabolic E vs. k band plots, and the interaction of electrons with the lattice in the neighborhood of the zone boundary has been mentioned. The above provide a substantial amount of insight into the understanding of the electronic structure of solids and its relationship to symmetry. It should not be forgotten, however, that band theory is a single electron theory with an approximate correction for correlation effects.

X.5 Problems

1. Show that the Bloch function labeled by **k** is a basis function for an irr. rep. of the translational subgroup, and that the Bloch function labeled by **K** + **k** is a basis function for the same irr. rep.
2. Show that the number of "particle in a box" states (X.24) at a given energy is given by the number of lattice points labeled by p_1, p_2, p_3 with a given value of $p = p_1^2 + p_2^2 + p_3^2$, and that the number increases like the volume increment of a sphere of radius $\sqrt{E}$. Show that the density of states, dN/dE for free electrons is given by

$$\frac{dN}{dE} = \frac{4\pi(2m)^{3/2}V}{h^3} E^{1/2} .$$

3. Using the free electron model calculate E_F, the energy of the highest filled state, for Na(s) (density = 0.97 g · cm^{-3}). Compare this energy with $\frac{3}{2}kT$.
4. Calculate the deBroglie wave length of a free electron at the Fermi level, as in problem 3, and compare with that of a wave function for such an electron.
5. Discuss briefly and qualitatively the effect of the lattice perturbation upon the density of states at the Brillouin zone boundary.
6. Use the fact that ∇E is a vector and the absence of vector invariants at special points (e.g., the zone center and some points on the zone boundary) to show that E vs **k** is at an extremum of these points.

References

1. Owens, J. P., Conard, B. R., Franzen, H. F.: Acta Cryst. **23**, 77 (1967)
2. Wiegers, G. A., Jellinek, F.: J. Solid State Chem. **1**, 519 (1970)
3. Heraldson, H., Kjekshus, A., Røst, E., Steffensen, A.: Acta Chem. Scand. **17**, 1283 (1963)
4. Slater, J. C.: J. Chem. Phys. **41**, 3199 (1964)
5. Goodenough, J. B.: in Proceed. Robert A. Welch Found. Conf. on Chem. Res. XIV Solid State Chemistry, (Milligan, W. O. (Ed.)) Houston, TX, 1971
6. Rundle, R. E.: in Intermetallic Compounds, (Westbrook, J. H. (Ed.)) John Wiley and Sons, Inc., NY, 1967
7. Altmann, S. L., Coulson, C. A., Hume-Rothery, W.: Proc. Roy. Soc., Ser. A 240, 145 (1957)
8. Engel, N.: Kem. Maanedsblad, **30** (5), 53; (6), 75; (8), 97; (9), 105; (10), 114, (1949)
9. Brewer, L.: in Electronic Structure and Bond Character in Relation to Electronic Structure and Crystal Structure, (Beck, P. A. (Ed.)) Interscience, NY, 1963
10. Pauling, L.: The Nature of the Chemical Bond, 3rd Ed., Cornell Univ. Press, Ithaca, NY (1960)
11. Rundle, R. E.: Acta Cryst. **1**, 180 (1948)
12. Bilz, H.: Z. Physik **153**, 388 (1958)
13. Neckel, A., Schwarz, K., Eibler, R., Weinberger, P., Rastl, P.: Ber. Bunsenges. **79**, 1053 (1975)
14. Nowotny, H.: Progr. Solid State Chem. **5**, 27 (1971)
15. Chevrel, R.: NATO Adv. Study Inst. Ser. B, B68 (1981)
16. Corbett, J. D.: Acc. Chem. Res., **14**, 239 (1981)
17. McCarley, R. E.: in Inorganic Chemistry: Toward the 21st Century, ACS Symp. Ser. No. 211, 1983
18. Simon, A.: Structure and Bonding **36**, 81 (1979)
19. Franzen, H. F.: Prog. in Solid State Chem. **12**, 1 (1978)
20. Franzen, H. F., Graham, J.: Z. für Krist. **123**, 2 (1966)
21. Adolphson, D. G., Corbett, J. D.: Inorg. Chem. **15**, 1820 (1976)
22. Marchiando, J. F., Harmon, B. N., Liu, S. H.: Physica, **99B**, 259 (1980)
23. Corbett, J. D., Marek, H. S.: Inorg. Chem. **22**, 3194 (1983)
24. Franzen, H. F., Smeggil, J., Conard, B. R.: Mat. Res. Bull., **2**, 1087 (1967)
25. Wells, A. F.: Structural Inorganic Chemistry, 4th Ed., Clarendon Press Oxford (1975)

26. Pearson, W. B.: A Handbook of Lattice Spacings and Structures of Metals and Alloys, Vol. 2, Pergamon Press, NY (1967)
27. Franzen, H. F.: in Titanium: Physico-Chemical Properties of its Compounds and Alloys, Atomic Energy Review, Special Issue No. 9, (Komarek, K. L. (Ed.)) Int. At. Energy Agency, Vienna, (1983)
28. Holmberg, B.: Acta Chem. Scand. **16**, 1245 (1962)
29. Franzen, H. F., Smeggil, J.: Acta Cryst. **B26**, 125 (1970)
30. Poeppelmeier, K. R., Corbett, J. D.: Inorg. Chem. **16**, 294 (1977)
31. Adolphson, D. G., Corbett, J. D.: Inorg. Chem. **15**, 1820 (1976)
32. Selte, K., Kjekshus, A.: Acta Chem. Scand. **17**, 2560 (1963)
33. Magneli, A.: Acta Cryst. **6**, 495 (1953)
34. Landau, L. D., Lifshitz, E. M.: Statistical Physics, Pergamon Press, London (1958)
35. Franzen, H. F., Merrick, J. A.: J. Solid State Chem. **33**, 371 (1980)
36. Popma, T. J. A., van Bruggen, C. V.: J. Inorg. Nucl. Chem. **31**, 73 (1969)
37. Heyding, R. D., Calvert, L. D.: Can. J. Chem. **35**, 449 (1957)
38. Franzen, H. F., Burger, T. J.: J. Chem. Phys. **49**, 2268 (1968)
39. Heine, V., McConnell, J. D. C.: Phys. Rev. Lett. **46**, 1092 (1981)
40. Neubuser, J., Wondratschek, H.: Maximal Subgroups of the Space-Groups, University of Karlsruhe, internal publication
41. Hahn, T. (Ed.): International Tables for X-ray Crystallography, Vol. A, International Union of Crystallography, D. Reidel Publishing Co., Boston (1983)
42. Haas, C.: Phys. Rev. **140** A863 (1965)
43. Franzen, H. F., Tuenge, R. T., Eyring, L.: J. Solid State Chem. **49**, 206 (1983)
44. Dismukes, J. P., White, J. G.: Inorg. Chem. **3**, 1220 (1964)
45. Hariharan, A. V., Powell, D. R., Jacobson, R. A., Franzen, H. F.: J. Solid State Chem. **36**, 148 (1981)
46. van Bruggen, C. F.: Investigations on Chromium Sulfides, Dissertation, University of Groningen (1969)
47. Jellinek, F.: Acta Cryst. **10**, 620 (1957)
48. Sundberg, M., Tilley, R. J. D.: J. Solid State Chem. **11**, 150 (1974)
49. Nguyen, T.-H., Franzen, H., Harmon, B. N.: J. Chem. Phys. **73**, 425 (1980)
50. Misemer, D. K., Nakahara, J. F.: J. Chem. Phys. **80**, 1964 (1984)
51. Nakahara, J., Franzen, H., Misemer, D. K.: J. Chem. Phys. **76**, 4080 (1982)
52. Mills, K. C.: Thermodynamic Data for Inorganic Sulphides, Selenides and Tellurides, Butterworth, London, 1974
53. Tuenge, R. T., Laabs, F., Franzen, H. F.: J. Chem. Phys. **65**, 2400 (1976)
54. Edwards, J. G., Franzen, H. F., Gilles, P. W.: J. Chem. Phys., **54**, 545 (1971)
55. Rossini, F. D., Wagman, D. D., Evans, W. H., Levine, S., Jaffe, I.: Selected Values of Chemical Thermodynamic Properties, Nat. Bur. of Stands. Circ. 500, U.S. Government, Washington, DC, 1952
56. Ziman, J. M.: Principles of the Theory of Solids, Cambridge University Press, (1972)

Subject Index